Learning Astronomy by Doing Astronomy

Collaborative Lecture Activities

SECOND EDITION

Learning Astronomy by Doing Astronomy

Collaborative Lecture Activities

SECOND EDITION

Stacy Palen

WEBER STATE UNIVERSITY

Ana M. Larson

UNIVERSITY OF WASHINGTON

W. W. NORTON & COMPANY

Independent Publishers Since 1923

W. W. Norton & Company has been independent since its founding in 1923, when William Warder Norton and Mary D. Herter Norton first published lectures delivered at the People's Institute, the adult education division of New York City's Cooper Union. The firm soon expanded its program beyond the Institute, publishing books by celebrated academics from America and abroad. By midcentury, the two major pillars of Norton's publishing program—trade books and college texts—were firmly established. In the 1950s, the Norton family transferred control of the company to its employees, and today—with a staff of five hundred and hundreds of trade, college, and professional titles published each year—W. W. Norton & Company stands as the largest and oldest publishing house owned wholly by its employees.

Editor: Erik Fahlgren
Project Editor: Laura Dragonette
Copy Editor: Laura Sewell
Editorial Assistant: Sara Bonacum
Managing Editor, College: Marian Johnson
Production Manager: Ashley Horna
Media Editor: Rob Bellinger
Associate Media Editor: Arielle Holstein
Media Project Editor: Danielle Belfiore
Media Editorial Assistant: Jasvir Singh
Managing Editor, College Digital Media: Kim Yi
Marketing Manager, Astronomy: Katie Sweeney
Design Director: Rubina Yeh
Designer: Juan Paolo Francisco
Director of College Permissions: Megan Schindel
College Permissions Manager: Bethany Salminen
Photo Editor: Travis Carr
Composition and Illustrations: Graphic World
Manufacturing: Transcontinental Printing

Permission to use copyrighted material is included in the credits section of this book.

ISBN: 978-0-393-69066-8

W. W. Norton & Company, Inc., 500 Fifth Avenue, New York, NY 10110
wwnorton.com

W. W. Norton & Company Ltd., 15 Carlisle Street, London W1D 3BS

1 2 3 4 5 6 7 8 9 0

Contents

About the Authors

STACY PALEN received her bachelor's degree from Rutgers University in 1993, and her PhD from the University of Iowa in 1998. Upon graduation, she spent four years as a post-doc/lecturer at the University of Washington, where she taught introductory astronomy 20 times in four years. This experience enabled her to really focus on students' conceptual difficulties with the course material and experiment with ways to address them. Stacy's astronomical research has focused on multi-wavelength studies of dying Sun-like stars. She has also conducted research in the teaching and learning of astronomy and served on multiple national committees that promote astronomy through education and public outreach.

Currently, Stacy is a full professor at Weber State University in Ogden, Utah. She lives on a small farm with her husband, two dogs, five dairy goats, five horses, and an ever-changing number of chickens.

ANA LARSON received a double bachelor's degree in physics and astronomy from the University of Washington in 1990 and a PhD in physics from the University of Victoria in 1996. Her thesis and subsequent ongoing research involve stellar properties and evolution. Upon graduation, Ana joined the University of Washington and is now a lecturer emeritus in astronomy. She has taught introductory astronomy approximately twice a year for the past 20-plus years, most frequently for the large lecture-based course, and has also taught advanced courses in astrophysics. Her development and thorough testing of introductory, lecture-related activities has been an ongoing effort for almost two decades. Having recognized that students need to see connections between what is read in the textbook, taught in class, and practiced in sections, she has created activities that are both relevant and challenging. Ana is currently teaching two introductory astronomy online courses at Seattle Central College and has incorporated the *Learning Astronomy by Doing Astronomy* activities into the courses. Students appreciate the pre-activity questions in preparation for assignments and the fact that they are actively engaged in their learning.

Preface

Dear Student,

The study of astronomy is filled with wonder, excitement, and surprises. While sitting and listening to someone lecture about wonder, excitement, and surprises can be very educational, it is much more fun to discover the surprises yourself, create your own excitement, and find your own answers to all the questions you wonder about. We have designed the activities in this book to do exactly that: to place in front of you the data, concepts, and skills, and then guide you as you develop your own understanding. Please be aware that there is an appendix in the back of the book with images in a larger format or in color; be sure to look there if you can't find a figure you need.

Throughout this book we emphasize a hands-on, minds-on approach in which you will be challenged to re-create discoveries, verify scientific principles, and question your preexisting notions. Sometimes this will be very challenging! It can be very hard to let go of a comfortable idea that you have held for a long time. It can be very hard to have the discipline to keep working, even when you don't always see how the project will end. But the investment of your time and effort will be worth it. You will be rewarded with a deeper understanding not only of the universe itself, but also of the methods astronomers use to explore the universe. You will begin to develop a feel for how astronomers ask questions, seek answers, and share results.

Each activity is designed and classroom-tested to illuminate a particular astronomical concept or principle. In some cases, such as discovering the expansion of the universe or a new planet around another star, you will be working with real data from real telescopes. In other cases, you will be using metaphors to explore inaccessible regions of space, like the area around a black hole. In every case, you will be required to engage with the material to develop the deepest understanding.

Experience is the best teacher. We have assembled these activities to give you real-world experience with the universe and the objects in it. As you work your way through the book, pause occasionally to marvel at how much we know about the universe and, perhaps more importantly, how much remains to be discovered and understood. We hope you will be inspired to ask new questions, collect your own data, and find your own answers.

Dear Instructor,

Studies of teaching and learning confirm that a hands-on approach is "hands-down" the best way to learn. When students actively engage with the material in a course, they make great strides in overcoming misconceptions, acquiring new knowledge, and building new skills. We have created this workbook by following the best practices of astronomical education research.

Each activity is built around a basic concept in astronomy and leads the student from a novice understanding to a deeper understanding through guided interaction. Students may interact with astronomical data or with common metaphors for astronomical objects, learning to predict the behavior and properties of astronomical objects. All of these activities have been "field-tested" in classrooms of various sizes, with students of varying backgrounds. All have been found to be effective for students whether they are learning in an in-person university setting, an open-enrollment college or university, or in an online course.

We have made a special effort to balance the activities among the four areas commonly found in introductory astronomy courses: physics and observations, the Solar System, stars, and galaxies and cosmology. You will find as many activities to use in the last half of your course as there are for the first half. Because of this broad coverage, the workbook may be used to accompany any textbook or it may be used as a stand-alone learning package for your students. In this edition, we have added nine new activities. Some are ranking or sorting activities, to help students organize their thoughts about astronomical processes, while others focus on current topics, to help students learn about and explore the cutting edge of science today. For example:

- Activity 13 introduces students to the underlying physical principles governing climate change, using only graphing skills.

- Activity 31 asks students to explore how gravitational waves reveal merging black holes.

- Activity 34 encourages students to sort the steps in the evolution of the universe.

We have included enough activities that you have some choice, but not so many as to overwhelm the students. With 36 activities from which to choose, you will have sufficient activities for a typical introductory course to use at least one per week.

These activities are designed to be used in the classroom, require no special equipment or preparation (although some would be enhanced by showing images on a screen), and can be completed within an hour by most students.

Students working in groups of two or three will make the most effective use of the activities, although it would also be possible for them to work alone. The sections of each activity are arranged in steps, guiding the students from initial knowledge-level questions or practice to a final evaluation or synthesis of what they have just accomplished. Students thus get practice thinking at higher cognitive levels.

Each activity has pre-activity and post-activity questions. The pre-activity questions have been designed to address some of the common misconceptions that students have, to relate familiar analogous terrestrial examples to the activity, and to act as a brief refresher in such things as scale factors, measurements, and basic mathematics review. These questions should act as mindsets for students as they transition from their other courses and activities to their learning of astronomy. The post-activity questions review the most important concepts introduced in the activity. Thus, the post-activity questions can be used as assessment tools, as reading quizzes, or to stimulate post-activity discussions. In the Second Edition, the pre- and post-activity questions are available in Smartwork5, Norton's online tutorial and homework system, so you can easily assess students' understanding before and after class. The pre- and post-activity questions continue to be available as PowerPoint slides to be used with any classroom response system.

We authors have been using activities like these in our classroom for more than 20 years. The Second Edition benefits not just from our experience but also from the experiences of adopters of the First Edition as well as reviewers. Based on this input, we have made improvements to the workbook, including the following:

- Nine activities—1, 2, 12, 13, 21, 24, 27, 31, and 34—are all new to the Second Edition.

- In each quantitative activity, we demonstrate the math logic needed and either show all of the solution or give students examples and ask them to complete just one or two simple calculations. Answers to similar calculations are shown in the activity—in a table, graph, or similar calculation—so students can learn to check their own work.

- Each activity is now self-contained. To make the workbook a resource that can truly be used stand-alone, we have provided a more complete introduction in the Step 1—Background section of each activity. We follow that information with a comprehension question that allows students to self-assess their understanding of the information provided in Step 1. Key terms are now listed at the end of the Learning Goals for each activity, and a complete glossary that includes these terms is now at the end of the workbook.

- We have standardized a final step we've called Putting It Together, which asks students to reflect on what they have learned and synthesize the information from the whole activity.

- We have included an appendix that contains some images in a larger format or in color. These color images were formerly available only via the online instructor's resources, which was inconvenient. Be sure to point out to students that they can tear out the assigned activity so they won't have to flip back and forth to see the activity and the referenced image simultaneously.

"Activity day" is always a favorite for us and students alike. We repeatedly find great joy in watching students figure out difficult concepts for themselves, develop confidence in their data analysis, graphing, and mathematical skills, and push each other to greater feats of intellectual bravery as they grow to understand that the universe is not magic. It is not impenetrable. It does not take a special talent or type of mind to understand it. It takes careful, thoughtful study and a willingness to ask every question that comes to mind and then chase down the answers. We are confident that you will see the same changes in your students that we see in ours. Please do check in with us and let us know how these activities work for you.

Acknowledgments

We would like to thank the instructors whose input at every stage improved the final product. We greatly appreciate their help.

Second Edition Reviewers

Jennifer Blue, Miami University
Liz Coelho, Pikes Peak Community College
Erin De Pree, St. Mary's College of Maryland
Jessica Ennis, Augsburg University
Caren Garrity, Hillsborough Community College
Idan Ginsburg, Tufts University
Carol Hood, California State University, San Bernardino
Jessica Lair, Eastern Kentucky University
Kaisa Young, Nicholls State University

First Edition Reviewers

Charles Kerton, Iowa State University
Nathan Miller, University of Wisconsin–Eau Claire
Dwight Russell, Baylor University
Don Terndrup, The Ohio State University

● ACTIVITY 1
The Celestial Sphere and Sky Maps

Learning Goals

In this activity, you will learn more about the features of the sky: where to look, when to look, how the sky changes, and why it changes. After completing this activity, you should be able to

1. contrast daily and yearly motion of the stars.

2. state the altitude of the north celestial pole (Polaris) from your location.

3. describe how stars move over the course of a year.

4. identify constellations that lie on the meridian and the ecliptic.

5. identify the locations of the solstices and equinoxes on the celestial sphere.

6. evaluate the effectiveness of viewing the night sky from different reference frames.

Key terms: celestial sphere, constellation, ecliptic, zodiac, celestial equator, equinox, solstice, zenith, horizon, meridian, latitude, north celestial pole (NCP), circumpolar, rotate, revolve

Step 1—Background

The **celestial sphere** is an imaginary sphere around Earth, with all the objects of the sky painted upon it. It's a useful idea for thinking about how the sky changes over the course of the year. Astronomers divide the entire sky into 88 **constellations**—distinct, irregularly shaped regions much like countries on Earth. The date, time, and location of an observation determine which constellations can be seen. Star maps show half of the sky as it appears on a particular date and time, and from a particular location.

The **ecliptic** is the path the Sun takes through the sky against the background stars. If the Sun is in the direction of Virgo, we might say, "The Sun is in Virgo." The set of constellations along the ecliptic make up the **zodiac**. The planets of our Solar System are always close to the ecliptic in the sky.

The **celestial equator** is directly above Earth's equator. Because Earth's pole is tilted with respect to its orbit, the celestial equator is tilted with respect to the ecliptic. The two locations in the sky where the celestial equator and the ecliptic cross are called **equinoxes**. This term is also used for the dates when the Sun is at those locations. For example, the northern vernal (spring) equinox is the position of the Sun on March 21, so March 21 is also known as the vernal equinox in the Northern Hemisphere. Solstices mark the locations on the ecliptic where the Sun is farthest from the equator, and also indicate the date when the Sun is at those locations. The star maps used in this activity show the sky at midnight on the dates for the northern vernal equinox, northern summer **solstice** (Sun is its farthest north of the celestial equator), northern autumnal equinox, and northern winter solstice (Sun is at its farthest south of the celestial equator).

At any time and in any location, the **zenith** is the point directly above your head. The **horizon** is a circle around you such that if you point at the zenith with one arm and the horizon with the other, your arms form a 90-degree angle. The **meridian** is the imaginary line in the sky that passes from due north to due south through the zenith. The star maps are completely accurate if we lived on the equator, with the celestial equator passing through our zenith and the north celestial pole on the northern horizon. The star maps used here have been modified slightly to show the approximate east-to-west horizon for 30 degrees and 45 degrees north **latitudes**, as well as the ecliptic.

We may never actually travel to the North Pole, but we can investigate the sky as it would look if we did stand there. If we were to look straight up—at our zenith—we would see the **north celestial pole** (NCP) directly overhead. The celestial equator wraps around our horizon, and all of the northern hemisphere stars are **circumpolar**—they neither rise nor set but are always above the horizon. We may see only half of the sky at the North Pole, but the stars we see never set.

Our emphasis will be on what is viewable at midnight as we look south on the dates of the equinoxes and solstices. We will be asking: Which constellations are rising in the east? Which ones are on the meridian? Which ones are setting in the west?

The Sun, Moon, and the stars rise in the east and set in the west, with an approximate 24-hour cycle. This motion is because Earth **rotates** on its axis—counterclockwise from west to east— as seen if we were looking down from above the North Pole, as sketched in **Figure 1.1**. Earth also **revolves** around the Sun in a counterclockwise direction. Another way of saying this is "Earth orbits the Sun counterclockwise." These two motions of Earth explain why the sky changes.

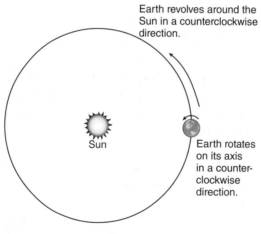

Earth revolves around the Sun in a counterclockwise direction.

Sun

Earth rotates on its axis in a counter-clockwise direction.

FIGURE 1.1

1. The locations on the celestial sphere where the ecliptic and celestial equator cross are called
   a. solstices   **b.** equinoxes   **c.** celestial poles   **d.** zeniths

---

Learning Astronomy by Doing Astronomy Second Edition

Step 2—Daily versus Yearly Motions of the Sky

The sketches in **Figure 1.2** indicate the motions you would see in the night sky looking north while using a camera and exposures of a few hours. East is to the right and west is to the left. Notice that one star, Polaris, stays in the same position in all three sketches. This is because Polaris is located very close to the north celestial pole (NCP).

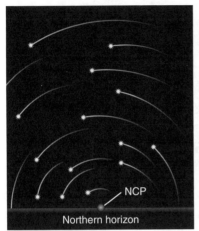

The NCP is on the northern horizon at the equator, latitude 0°.

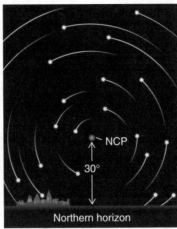

The NCP is 30° above the northern horizon at latitude 30°N.

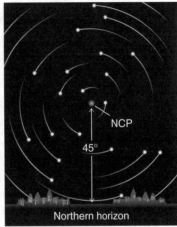

The NCP is 45° above the northern horizon at latitude 45°N.

FIGURE 1.2

2. There is definitely a trend in the altitude of the NCP above the northern horizon as viewed from different northern latitudes. State that trend.

3. At what latitude are you currently? latitude _____

4. If tonight were clear and you went out to find the star Polaris, how many degrees would it be above the northern horizon? _____

5. Is the motion of the stars around Polaris (the NCP) due to the _____ **rotation** (daily motion) or _____ **revolution** (yearly motion) of Earth?

   As Earth revolves around the Sun, the apparent position of the Sun against the background constellations changes. As **Figure 1.3** shows, on September 1 the Sun is seen in the direction of Leo. At midnight on September 1, we would look in the direction of the constellation Aquarius. By December 1, after 3 months have gone by, Earth has traveled far enough in its orbit that the Sun is in Scorpius. At midnight on that date, we would look into the direction of Taurus. Note the location of Earth with respect to the Sun at the equinoxes and solstices.

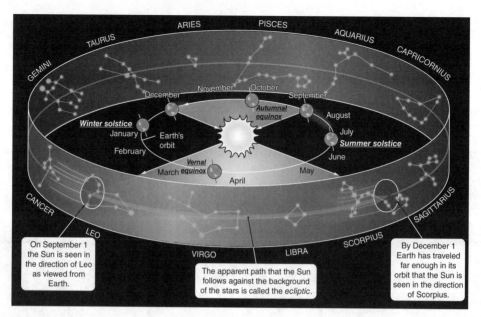

FIGURE 1.3

6. If we assume the 12 constellations of the zodiac were evenly distributed along the ecliptic, about how long would the Sun "stay" in each of those constellations? _____

7. Imagine that at noon on the northern summer solstice there is a total eclipse of the Sun, and you can see the background stars in the middle of the day. The sky might look something like the sketch in **Figure 1.4**. Notice that the Sun is in the constellation of Gemini. On the same day 6 hours later, in which constellation would the Sun be?

   a. Cancer   **b.** Gemini   **c.** Taurus   **d.** Orion

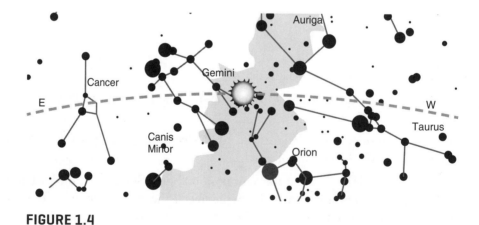

FIGURE 1.4

Step 3—Reading Star Maps

We saw how the background of stars and the constellations slowly changes over the course of a year in Figure 1.3. While objects rise and set on a daily basis, the time of rising and setting will not be the same over a year. The star maps shown in **Figures 1.5–1.8** in the appendix will help you visualize what the sky looks like at four instances during a year. Since these are drawn from the perspective of a viewer looking south, west is on the right and east is on the left.

8. Carefully study **Figure 1.5**, which shows the night sky at midnight on the vernal equinox, the first day of spring. Use a straight edge and draw in the meridian. List four constellations that are on or close to the meridian (and are above the southern horizon for your location) at around midnight on the vernal equinox.

9. This is also the sky at noon on September 21; that is, it is a map of the daytime sky on the autumnal equinox. In which constellation is the Sun located on the autumnal equinox?

   a. Libra   **b.** Virgo   **c.** Leo

10. Draw in the meridian on **Figure 1.6**, as you did for Figure 1.5. This figure shows the night sky at midnight on the first day of summer, the summer solstice. List four constellations that are on the meridian and above the horizon at your latitude.

11. Pick one of the constellations that was on the meridian at midnight on the spring equinox (from Figure 1.5), name it, and describe how its location in the sky changed over one season.

12. The star map of Figure 1.6 also depicts the sky at noon on the winter solstice. Is the Sun _____ **above** or _____ **below** the celestial equator in the winter?

13. In which constellation is the Sun located on December 21?

    a. Capricornus   **b.** Sagittarius   **c.** Libra

14. As you take a look at **Figure 1.7**, which shows the night sky at midnight on the autumnal equinox, note that the bright star in Lyra, Vega, that was on the meridian on the summer solstice, is now setting in the northwest. Approximately how much time has passed between Figure 1.6 and Figure 1.7?

15. How does that amount of time relate to Vega's change in its location in the night sky?

16. This map shows the stars in the sky at noon on March 21. In which constellation is the Sun on March 21?

    a. Pegasus   **b.** Pisces   **c.** Aquarius

17. List five constellations shown in **Figure 1.8** that lie on the ecliptic. This figure shows the night sky at midnight on the first day of northern winter—the winter solstice.

     Learning Astronomy by Doing Astronomy Second Edition

18. This star map also represents the sky at noon on June 21. Is the Sun located _____ **above** or _____ **below** the celestial equator in the summer?

19. In which constellation is the Sun on June 21?

    a. Cancer   **b.** Gemini   **c.** Taurus

Step 4—Putting It Together

20. Return to Figure 1.3 and draw straight lines from Earth at each equinox and solstice through the Sun to the opposite constellation. Then, review your answers given in Step 3 for Figures 1.5–1.8. How does the way we view the yearly motion of Earth in Figure 1.3 relate to how we view the night sky using the star maps of Figures 1.5–1.8? Was investigating the sky from these two points of view helpful or not? Explain.

21. Start with listing at least three "points" (topics, concepts, visualizations) that this activity has covered. Then write an essay of around 200–300 words that summarizes what you have learned about those points. You should be able to use at least five of the key terms.

● ACTIVITY 2
Designing a Scale Model of the Solar System

Learning Goals

In this activity, you will learn how to create a scale model of the Solar System. After finishing this activity, you will be able to

1. calculate the scale factor for the relative sizes and distances used in the model.

2. apply the scale factor to missing quantities.

3. locate the eight major planets along the model Solar System route.

4. summarize the steps used in creating a scale model.

5. use the metric system and scientific notation.

Key terms: scale, scaling, scale factor, metric, units, astronomical unit (AU), scientific notation

Step 1—Background

You may be familiar with the concept of **scale** from scale models. For example, you could purchase a scale model of a 2017 Ford F-150 Raptor Pickup for under $25. Its scale is 1/24, meaning that the real thing is 24 times as large as the model. Every inch on the model corresponds to 24 inches on the real truck. Similarly, a toy model car that is scaled 1/32 means that every inch on the model equals 32 inches on the real car, or it could mean 1 millimeter on the model equals 32 millimeters on the real car.

You are likely familiar with maps, which also use the concept of **scaling**. Have you ever looked at a map and noticed that 1 inch on the map equals 5 miles on land? That is scaling, and the **scale factor** is 1 inch = 5 miles. In this activity, you will be making a "map" of the Solar System, using scale factors to shrink the Solar System down to human scale—about the size of Washington, D.C. This will help you to understand the relative sizes and distances in the Solar System.

For astronomy (and most sciences) we use the **metric** system of **units**, which is based on the meter (one meter is the same length as three feet plus three inches). The metric system is convenient because conversions from one unit to another are always factors of 10; this is called a "base 10" system. Working with a base of 10 makes calculations and conversions so much easier: 10 millimeters = 1 centimeter; 100 centimeters or 1,000 millimeters = 1 meter; 1,000 meters = 1 kilometer. Converting back and forth from meters to kilometers simply means moving the decimal point to the left or right.

In this activity, we will use two different scales: one for the diameters of planets, and one for their distances from the Sun. For the diameters of planets, the scale factor will relate millimeters on Earth to kilometers in space. Because distances between planets are so much larger than the sizes of planets, we need a different scale factor for distances. For the distances of planets from the Sun, the scale factor will relate meters on Earth to kilometers in space.

As an example, let's assume we scale the planets' diameters down by a factor of 500. We could have used 1 mm on our model planet equals 500 km on a real planet. This means the scale factor is 500 km per mm, which is more compactly written as 500 km/mm. The Earth has a diameter of about 12,700 km. How large would our model Earth be on this Solar System scale model?

$$\frac{12{,}700 \text{ km}}{500 \text{ km/mm}} = 25.4 \text{ mm}$$

For the distances, let's assume we use a scale factor of 1 meter equals 500,000 kilometers. So, every meter on the ground equals 500,000 kilometers in space. The Earth's average distance from the Sun in space is 1.496×10^8 km. This number is so useful that we give it its own name: one **astronomical unit (AU)**. The average distance of the Earth from the Sun is 1 AU = 1.496×10^8 km. If we apply the scale factor to this distance, then the scaled-down distance from Earth to the Sun would be:

$$\frac{149{,}600{,}000 \text{ km}}{500{,}000 \text{ km/m}} = 299 \text{ m}$$

In other words, if you shrank the Solar System down using these scale factors, Earth would be a ball 25.4 mm in diameter, located 299 meters away from the center (the scaled-down Sun) of the scaled-down Solar System.

Notice that we have a lot of zeroes in the equation above. Typing those zeroes into your calculator is tedious. Because we will be talking about large distances, with lots of zeroes, we will be using **scientific notation** in this activity. Scientific notation is a way of making numbers more compact. To put a large number into scientific notation, count up all the digits except the first one. This number of digits will be expressed as an exponent (sometimes called "a power") of 10. For example, 149,600,000 km has eight digits after the first one. Using scientific notation, you could write it as 1.496×10^8 km. Your calculator has special buttons (sometimes labeled "EE" or "ee") for handling scientific notation, saving you the trouble of having to carefully count and type in all the zeroes. If you take a moment to figure out how to input scientific notation in your calculator, you will save a lot of time throughout this course.

1. Assume that instead of 500 km per mm for the size of the Earth, the scale factor was 790 km/mm. Using this scale, what would the scaled-down size of the Earth be?

_____ mm

Step 2—Calculating and Applying the Scale Factor for the Model

To scale the real Solar System down to a model Solar System that fits within a city, we need to determine the two scale factors we will use to reduce both the diameters and the distances. For the diameters, that means how many kilometers in the real Solar System are represented by a millimeter on Earth. For the distances from the Sun, which are much larger than the diameters, this scale factor will be how many kilometers a meter represents on the map.

Here, we have chosen to make the Sun 2,200 mm (2.2 meters) in size. This is a choice we made that makes ALL the scaled-down objects large enough to see, but small enough to hold. From this choice, you can calculate the scale factor:

2. Divide the actual diameter of the Sun (1.39×10^6 km) by the scaled-down diameter (2,200 mm) to find the scale factor for diameters:

_____ km/mm

3. On this scale, Jupiter's scaled-down diameter is 226.3 mm. Use the diameter of Jupiter from **Table 2.1** to double-check your calculation of the scale factor. You should get the same answer that you found for question 2.

_____ km/mm

4. Use that scale factor to fill in the missing scaled diameters (in millimeters) in Table 2.1. For example, if you were finding the scaled diameter of Saturn, divide the diameter of Saturn in kilometers by the scale factor:

$$\frac{1.15 \times 10^5 \text{ km}}{631.8 \text{ km/mm}} = 182.0 \text{ mm}$$

○ TABLE 2.1

Actual and scaled values for the diameters and orbit distances for the Sun and planets

OBJECT	DIAMETER (KM)	SCALED DIAMETER (MM)	ORBIT DISTANCE (KM)	SCALED ORBIT DISTANCE (M)	ORBIT DISTANCE (AU)
Sun	1.39×10^6	2,200			
Mercury	4.87×10^3		5.80×10^7	91	0.39
Venus	1.21×10^4	19.2	1.10×10^8		0.72
Earth	1.27×10^4		1.50×10^8	236	1.00
Mars	6.76×10^3		2.30×10^8	363	1.52
Jupiter	1.43×10^5	226.3	7.80×10^8		5.20
Saturn	1.15×10^5	182.0	1.40×10^9	2,207	9.54
Uranus	4.69×10^4	74.2	2.90×10^9		19.18
Neptune	4.54×10^4		4.50×10^9	7,093	30.06

5. The scale factor for orbital distances will be different, because distances are so much larger than diameters in the Solar System. Use the data for Mercury to find the scale factor for the orbit distances (similar to what you did for question 2). The units for this scale factor will be kilometers per meter.

_____ km/m

6. Double check your calculation of the scale factor using the data for Earth (similar to what you did for Jupiter for question 3). **Note:** Answers will not be exact and will differ slightly because of rounding.

_____ km/m

7. Use this orbital distances scale factor to fill in Table 2.1 with the missing scaled orbit distances values (in meters). For example, to find the scaled orbital distance for Saturn, divide the orbital distance to Saturn by the scale factor:

$$\frac{1.4 \times 10^9 \text{ km}}{6.34 \times 10^5 \text{ km/m}} = 2{,}208 \text{ m}$$

8. Summarize the process you used for questions 2–7, addressing the values you used, how you obtained each of the two scale factors, and why the scale factor for the orbit distances is about 1,000 times larger than that used for scaling the diameters. Your summary should be detailed enough that a classmate could read it and repeat the exercise to get the same results as you did. Be sure to use many of the key terms from this activity.

Step 3—Locating the Planets along the Model Solar System Path

Now that the calculations are done, imagine that you are in charge of creating a Solar System walk in Washington, D.C. You will place nine balls with the scaled-down diameters you have calculated (one for each planet, and one for the Sun) at the appropriate distances within a map of Washington, D.C. Place the 2.2-meter (2,200 mm, or about 7 feet 3 inches diameter) Sun in the middle of the Robert F. Kennedy Memorial Stadium. The planets will be placed at their scaled distances from the model Sun along a straight path (as much as possible). Notice as you place the planets that visitors to our model Solar System will walk through Lincoln Park, pass by the Supreme Court, walk *through* the Capitol building, pass the various Smithsonian museums and galleries, the White House (to the north), the World War II Memorial, and the Vietnam Veterans Memorial, and end up close to the Lincoln Memorial. If visitors continue on past Neptune to the Lincoln Memorial, they will have walked 7,500 meters, or about 4.7 miles in total.

9. Placing **Figure 2.1** and **Figure 2.2** side by side would form a long rectangular map of Washington, D.C. Mark and label the locations of the eight planets on the maps given in Figure 2.1 and Figure 2.2 using the data for the scaled distances in Table 2.1. The flags along the route mark the distance of that location from the model Sun located in the stadium. Since the first four planets—Mercury, Venus, Earth, and Mars—are all less than 500 meters from the Sun at this scale, you will have to mark the locations of those planets in between the Sun and the first flag.

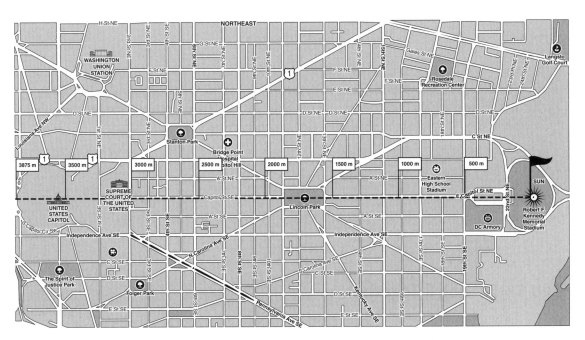

FIGURE 2.1

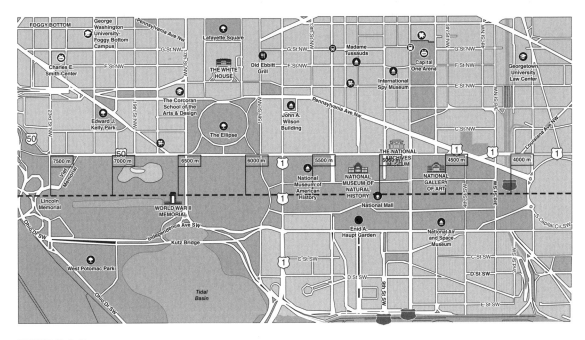

FIGURE 2.2

Step 4—Analysis of the Scale Factor Chosen

10. Why was the scale factor for planet diameter chosen to make Mercury so small, less than 8 mm? In other words, why not make Mercury four times the diameter used here so some surface features could be seen while also making the distances between the Sun and its planets four times as large?

11. Study the column that lists the actual distances of the orbits in astronomical units (AU). By definition, the Earth's distance from the Sun is 1 AU. Are the other distances compared to that of the Earth easier to visualize when stated in AUs? Why or why not?

Step 5—Putting It Together

12. Create a promotional advertisement for the new Solar System walk to be located in Washington, D.C. Incorporate what you have learned about the distances between planets and their relative sizes; use at least three key terms from this activity. Emphasize the process you used to create the walk, including what the scale factor is and how it was determined. What public benefits will this Solar System walk provide?

● ACTIVITY 3
Where on Earth Are You?

Learning Goals

In this activity, you will learn about the coordinate systems that are used on Earth and how our location on Earth and Earth's orbit around the Sun are related to the seasons. You should also be able to

1. recognize that the Sun and stars appear differently at different locations on Earth.

2. summarize how these differences lead to seasons on Earth.

3. state where the seasons are most and least extreme on Earth and how this difference follows from the location of the Sun in the sky.

Key terms: maximum altitude, summer solstice, winter solstice, autumnal equinox, vernal equinox, Arctic Circle, Tropic of Cancer, Tropic of Capricorn, equator, zenith, Antarctic Circle, celestial sphere, celestial poles, celestial equator, ecliptic, circumpolar

Step 1—Background

Every day, the Sun reaches a **maximum altitude** above the horizon around midday, when the Sun is highest in the sky. This altitude depends on both the observer's location on Earth and the time of year. The maximum altitude of the Sun at local noon changes over the course of a year because the Earth's rotation axis is tilted by 23.5 degrees with respect to the plane of its orbit around the Sun. Four special dates track the annual movement of the Sun through the sky. The date when the Sun is highest in the sky is the **summer solstice** (around June 21 in the Northern Hemisphere). This is the longest day of the year; on this day, the Sun is above the horizon longer than on any other day. The **winter solstice** (around December 21 in the Northern Hemisphere) marks the date when the Sun is lowest. This is the shortest day of the year; the Sun is above the horizon for less time than on any other day. There are two dates when the days and nights are equal—roughly 12 hours each—and those are called the equinoxes. The equinox in the fall is called the **autumnal equinox** (around September 21 in the Northern Hemisphere), and the equinox in the spring is called the **vernal equinox** (around March 21 in the Northern Hemisphere).

There are five identified major circles of latitude on Earth linked to locations of the Sun at certain times during the year. The **Arctic Circle** is at 66.5 degrees north latitude. For those living at that latitude, the Sun is found at the southern horizon at noon on the winter solstice (around December 21 each year). The **Tropic of Cancer** circles the Earth at 23.5 degrees north latitude; the **Tropic of Capricorn**, at 23.5 degrees south latitude. The **equator** lies right in between at latitude 0 degrees. If you lived on the equator, the Sun would be at your **zenith** (straight overhead) at noon on an equinox. While there are approximately 4 million people living above the Arctic Circle, the **Antarctic Circle**, located at 66.5 degrees south latitude, cuts through just the Antarctic Peninsula and the research bases there. An estimated 4,000 people reside there during the southern summer; only 1,000 during the winter.

It is sometimes useful to pretend that the sky is a giant sphere centered on Earth, called the **celestial sphere**. The points on the celestial sphere right above the North and South poles are called the north and south **celestial poles**. The **celestial equator** is a circle on the celestial sphere located directly above Earth's equator. The **ecliptic** is the path of the Sun through the sky over the year. This path is another circle, tilted with respect to the celestial equator. The ecliptic and the celestial equator cross at two places. On an equinox, the Sun will be at one of these two crossing points. Therefore, these two locations are also called the vernal and autumnal equinox; the term is used for *both* the location in the sky and the date on which the Sun is at that location. Similarly, the summer and winter solstice are terms used for both a date and a location in the sky.

If you lived at either the North or South pole, you would see only one-half of the celestial sphere, and all of the stars you could see would never rise or set; they all would be **circumpolar**. If you lived at the equator, all the stars would rise and set each day (although some would be up during the daytime, and you couldn't see them). No stars are circumpolar if you live at the equator.

1. How many times each year does the Sun pass through your zenith if you live on the equator?

_____ times

Step 2—The Celestial Sphere

A simplified view of the imaginary celestial sphere is given in **Figure 3.1**. Earth is at the center, and the shaded plane is the plane of the Earth's orbit. The circle on the celestial sphere where this plane crosses is the ecliptic. From our point of view, the ecliptic becomes the path of the Sun along the celestial sphere through the year.

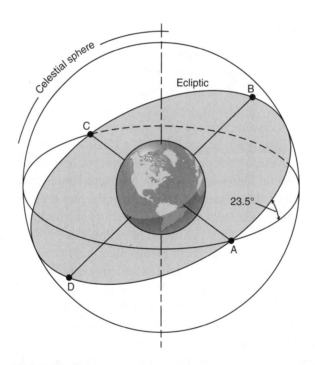

FIGURE 3.1

2. In Figure 3.1, notice the four special points (A–D) marked along the ecliptic. Use the perspective of an observer in the Northern Hemisphere and identify the letter that corresponds with the following positions of the Sun on the celestial sphere:
- _____ summer solstice
- _____ winter solstice
- _____ vernal (spring) equinox
- _____ autumnal (fall) equinox

3. Clearly label the north and south celestial poles and the celestial equator on Figure 3.1.

Step 3—Coordinates on Earth

We next examine Earth at two points during its orbit, as shown in **Figure 3.2**. One of these points is northern winter solstice, and the other is northern summer solstice. Sunlight is coming from between the two images. The Sun is so far away that all the rays from the Sun are parallel to the short side of this page.

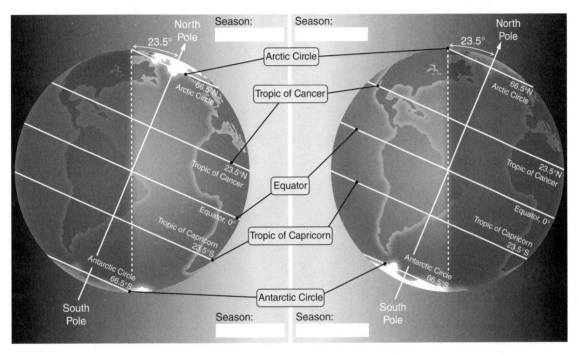

FIGURE 3.2

4. On Figure 3.2, fill in the four blanks beneath "Season" using the labels *summer* or *winter* where it applies to either the Northern or Southern hemispheres.

5. For an observer in the Northern Hemisphere, describe in general terms how the maximum altitude of the Sun changes throughout the year.

6. Considering the location of the Sun in the sky through a year, where on Earth would you expect to find the greatest range of temperatures? Why?

7. The five major circles of latitude are labeled on Figure 3.2. Review the background information in this activity that pertains to these circles of latitude. Consider what is unique among them and what observers of the Sun would see from each location. Fill in **Table 3.1** by matching the observations with the locations. More than one location may apply to each observation.

⚪ TABLE 3.1

Descriptions of the sky at various locations on Earth

DESCRIPTION OF SKY	LOCATION ON EARTH
_____ Sun is seen at the zenith twice during the year.	a. At the North Pole
_____ North circumpolar stars are seen.	b. At the Tropic of Cancer
_____ Sun is at the zenith only once during the year.	c. Everywhere south of the Tropic of Cancer
_____ North celestial pole is seen at the zenith.	d. Everywhere north of the equator
_____ All stars rise and set.	e. At the equator
_____ All northern stars are circumpolar.	f. Everywhere south of the equator
_____ Celestial poles are seen on the horizon.	g. Everywhere north of the Tropic of Capricorn
_____ South celestial pole is seen at the zenith.	h. At the Tropic of Capricorn
_____ All southern stars are circumpolar.	i. At the South Pole
_____ Sun is directly overhead at local noon on June 21.	
_____ Sun is directly overhead at local noon on December 21.	

Step 4—Putting It Together

8. No matter where you live on Earth, during the course of the year the Sun moves through an angle of 47 degrees between the solstices. For the Northern Hemisphere, this means from its highest point in the sky at local noon on the summer solstice to the lowest point in the sky at local noon on the winter solstice. This angle is, of course, just twice the angle of the tilt of the Earth's rotation axis of 23.5 degrees. Using the knowledge you gained in this activity, explain how the tilt of Earth's axis leads to our experiencing seasons on Earth. Use at least three key terms in your explanation.

● ACTIVITY 4
Studying the Phases of the Moon

Learning Goals

Understanding the phases of the Moon requires visualizing the Earth-Moon-Sun system in three dimensions. In this activity, you will develop this ability by learning how to

1. successfully replicate the motions of the Earth and Moon, as well as their positions with respect to the Sun at each lunar phase.

2. explain the continuity of the Moon phases worldwide.

3. use an Earth-Moon figure to disprove a common misconception that Moon phases are caused by Earth's shadow.

4. correctly order the phases of the Moon.

Key terms: rotate, revolve, new Moon, waxing crescent, first quarter Moon, waxing gibbous, full Moon, waning gibbous, third quarter Moon, waning crescent

Step 1—Background

The Earth, Moon, and Sun form an interconnected, moving system. The Earth **rotates** on its axis while it **revolves** around the Sun, which is also rotating. The Moon rotates on its axis, once for each time it revolves around the Earth. You may better understand the phrase "revolves around" if you replace it with "orbits." Stating "The Moon rotates around the Earth" or "The Earth rotates around the Sun" is incorrect.

Perhaps you have been taught about the phases of the Moon and how those phases depend on the Earth-Moon-Sun relative positions and our view of the Moon at each of those positions. Maybe you had to memorize the names of the phases—**new Moon, waxing crescent, first quarter Moon, waxing gibbous, full Moon, waning gibbous, third quarter Moon, waning crescent**. This activity should lead you to a better understanding of the cause of the phases of the Moon.

1. Complete the following statements using *revolves* or *rotates*.

- The Earth _____ once every 24 hours.

- The Moon _____ once on its axis for every revolution around Earth.

- One year is the amount of time it takes Earth to _____ around the Sun.

- The Moon _____ around the Earth in approximately 29.5 days.

Step 2—The Continuity of the Phases of the Moon

You will use a model of the Earth-Moon system to explore the phases of the Moon as they are seen on Earth every day. Take a full sheet of paper and crush it into a ball to represent Earth. Next, take one-quarter of a sheet of paper and crush it into a smaller ball to represent the Moon. You will start with a new Moon, the same geometry as shown in **Figure 4.1**. The Moon moves nearly 12° in its orbit every 24 hours (360° divided by a rounded-off 30 days in a month). For simplicity, consider all locations to mean those for viewers in the Northern Hemisphere—those who see the Moon located in the southern part of the sky as it crosses the meridian.

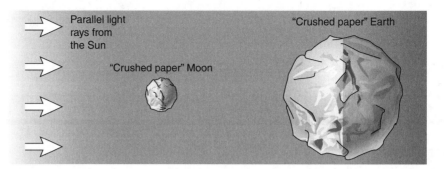

FIGURE 4.1

Place the models of the Earth and the Moon on **Figure 4.2**, found in the appendix. Imagine the sunlight coming from the front of the classroom. Orient your paper so that the "Day 0 & 30" location on the graph is toward the "Sun." Set Earth on the center of the polar graph, and then place the Moon around the edge of the polar graph in the position of new Moon. Make Earth rotate through 1 day while moving the Moon in its orbit (revolving) 12° during each day.

2. Starting with a new Moon, rotate the paper Earth once. For this model, do all locations on Earth see approximately a new Moon (that is, if we could actually "see" a new Moon)?

Circle **Yes** or **No**

3. Each time Earth rotates, 1 day passes, and the Moon moves 12°. Move the Moon counter-clockwise in its orbit, and rotate Earth once. Repeat this process (move the Moon 12° in its orbit, and rotate Earth once) until first quarter Moon (7.5 rotations of Earth). Do all locations on Earth see close to a first quarter Moon, over the course of one rotation of Earth?

Circle **Yes** or **No**

4. Move the Moon and rotate Earth until the alignment is Sun-Earth-Moon (Moon on the "Day 15" mark). What is the phase of the Moon? _____

5. Do all locations on Earth see this phase, or very close to this phase, over the course of one rotation of Earth?

Circle **Yes** or **No**

6. Continue another 7.5 days. What is the phase of the Moon? _____

7. Do all locations on Earth see close to this phase, over the course of one rotation of Earth?

Circle **Yes** or **No**

8. Briefly summarize your answers for questions 1–7, explaining how the phases of the Moon are similar for all locations on Earth over the course of one Earth rotation, throughout the time the Moon revolves. By doing so, you will also show that the phases of the Moon are continuous.

Step 3—Phases and Earth's Shadow

A common misconception is that the phases of the Moon are caused by Earth's shadow. **Figure 4.3** depicts the orbit of the Moon around Earth. The orbit is shown in perspective, from an angle between edge-on and directly above the orbit. The Sun is far off the paper to the left. Earth's shadow is shown to scale. The plane of the Moon's orbit is tilted with respect to the plane of Earth's orbit. Twice per year, the Moon is in the plane of Earth's orbit AT THE SAME TIME that it is in new or full phase. It is at these times that eclipses can happen. Otherwise, eclipses cannot occur because the shadow of Earth is above or below the Moon.

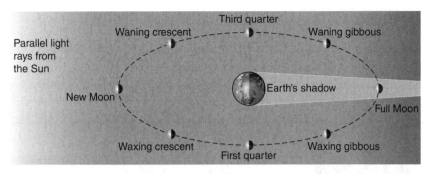

FIGURE 4.3

9. First quarter Moon is shown in **Figure 4.4**. Draw the shadows of the Earth and Moon. Use the shape of Earth's shadow in Figure 4.3 as a guide.

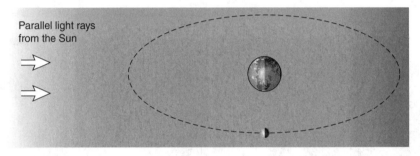

FIGURE 4.4

10. A photograph of a first quarter Moon is shown in **Figure 4.5**. Where is the Sun in this image? Where is Earth? Is there any way for Earth's shadow to darken the left side of the Moon (from our point of view) during its first quarter? Explain, taking into consideration the position of the Sun and the Earth relative to the Moon in this photograph.

FIGURE 4.5

11. In science, a hypothesis can be disproved by only one actual observation that shows it is wrong. Combine your work with Figure 4.4 and the observation of Figure 4.5 and state how it is impossible for the phases of the Moon to be caused by Earth's shadow.

Step 4—Understanding the Order of Moon Phases

12. In **Figure 4.6,** the phases of the Moon are shown in random order. Starting with "k," which represents a new Moon, sort the phases in the order they would occur, from new Moon to full Moon and back to new again. Write the corresponding letters in order below.

k											k

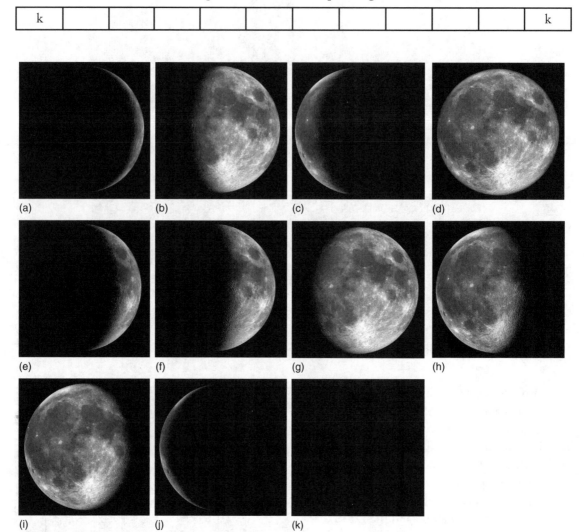

FIGURE 4.6

Step 5—Putting It Together

13. Summarize what you have learned about the phases of the Moon, using at least three of the key terms introduced in the learning goals.

Name _____ Date _____ Section_____

● ACTIVITY 5
Working with Kepler's Laws

Learning Goals

In this activity, you will learn about Kepler's geometric model of planetary orbits and, upon completion, will be able to

1. determine the properties of an ellipse through Kepler's first law.

2. apply these properties to planetary orbits using Kepler's second law.

3. understand what a universal law is, and explain why Kepler's third law is a universal law.

Key terms: ellipse, minor axis, major axis, foci, focus, perihelion, aphelion, period, semimajor axis, astronomical unit (AU), eccentricity, universal law

Step 1—Background

Johannes Kepler was a persnickety individual when it involved understanding the data given to him by Tycho Brahe. Kepler took a particularly precise, careful, even fussy approach to his calculations of the planetary orbits. By adopting the shape of an **ellipse** for the orbits of the planets, his predictions precisely fit the observations of their positions at any given time. Ellipses look like elongated circles, as seen in **Figure 5.1**. An ellipse will have a short axis, known as the **minor axis**, and a long axis, known as the **major axis**. An ellipse is drawn around two points or **foci**. For an elliptical orbit, the Sun is located at one **focus**; there is nothing at the other focus. This observation, that planets move in ellipses, is Kepler's first law. Kepler was also able to describe the motions of the planets based on the ellipse. Planets moved the fastest when they were passing closest to the Sun, or at **perihelion**. Planets moved the slowest when they were the farthest from the Sun, or at **aphelion**. Kepler's second law describes how the speed of the planet changes with position in the orbit. Years later, after studying the data for the planets known at that time, Kepler found that there was a definite relationship between how long a planet took to orbit the Sun—its **period**—and its average distance from the Sun. The average distance from the Sun is equal to the **semimajor axis**, which is found by dividing the length of the major axis by two. This relationship is Kepler's third law. We will work with all three of Kepler's laws in this activity.

1. Highlight the semimajor axis of the ellipse in Figure 5.1. If the units are **astronomical units** (AU, average distance of the Earth from the Sun), what is the semimajor axis of the ellipse?

_____ 3 ___ AU

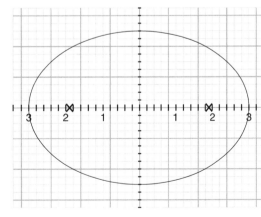

FIGURE 5.1

Step 2—Kepler's First Law

Figure 5.1 features Kepler's first law: The orbits of the planets are ellipses with the Sun at one of the foci.

2. Clearly label the following parts of an ellipse on Figure 5.1:

   a. focus (there are two)   **b.** semimajor axis

   c. minor axis   **d.** center

3. **Eccentricity** is a measure of the degree of "flattening" of the ellipse. The eccentricity of an ellipse is defined as the distance from a focus to the center of the ellipse divided by the length of the semimajor axis. You already found the length of the semimajor axis in question 1. Use the graph paper grid to find the distance from the focus to the center for the ellipse in Figure 5.1.

   Distance from focus to the center = _____

4. Calculate the eccentricity of the ellipse in Figure 5.1 by dividing the distance from the focus to the center by the semimajor axis.

   Eccentricity = _____

5. A circle is a special ellipse, one with both foci at the same point. The eccentricity of a circle is 0. The value of the eccentricity of an orbit may run from 0 to almost 1. **Figure 5.2** shows four examples of ellipses. Rank them from the smallest eccentricity to the largest.

   Smallest _____ _____ _____ _____ Largest

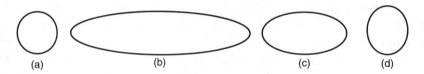

FIGURE 5.2

6. State how to determine the eccentricity of an ellipse. If you'd like, you may draw a figure or figures to express your thoughts.

Step 3—Kepler's Second Law

Kepler's second law states that as a planet moves around its orbit it sweeps out equal areas in equal times. This law can be difficult to visualize; it may be easier to consider this simpler version:

A planet travels faster when nearer to the Sun and slower when farther from the Sun.

7. Examine **Figure 5.3**, which shows a planet, D, orbiting the Sun, A, in a counterclockwise direction. Match the following terms with the letter that identifies the location in the figure:

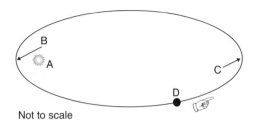

- focus _A_
- aphelion _C_
- perihelion _B_
- increasing speed _____ to _____
- decreasing speed _D_ to _C_
- planet has fastest speed _____
- planet has slowest speed _____

Not to scale

FIGURE 5.3

8. State Kepler's second law and give the physical reason why planets behave this way.

Step 4—Kepler's Third Law

Kepler's third law relates the time it takes a planet to go around the Sun (the period, P) to the semimajor axis, A, of the orbit. If we measure P in years and A in astronomical units (AU), the simplified relationship (formula) is:

$$P^2 = A^3$$

9. According to Kepler's third law, all orbits with the same semimajor axis have the same period. The two orbits in **Figure 5.4** have the same value for A and thus must have the same P, but how is this possible when the orbits have such different eccentricities? (**Hint**: With a circular orbit, $e = 0$, does the distance of an object orbiting the Sun change?)

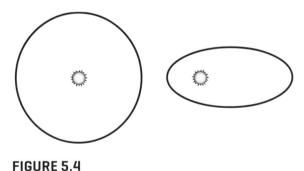

FIGURE 5.4

We know that Kepler's laws work for all objects that orbit the Sun, and we assume that the orbits of the comets, asteroids, and moons obey the first two laws, but what about Kepler's third law? Is it a **universal law**, one that is applicable literally across the universe? If it is, then we should find that $P^2 = A^3$ applies in all cases.

The period versus distance from Jupiter for the Galilean moons is shown in **Figure 5.5a**. **Figure 5.5b** shows the period versus distance from the star Trappist-1 for its six planets. The star Trappist-1 is 39.46 light-years away. Its planetary system was discovered in 2016, and it is just

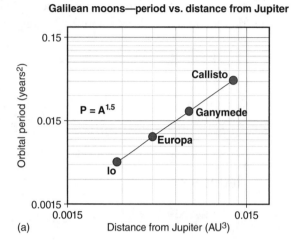

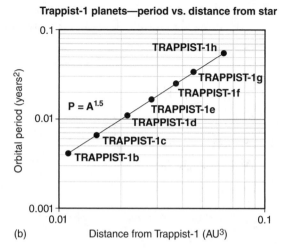

FIGURE 5.5

one of a large number of other stars recently discovered as having multiple planets orbiting it. The data for both systems are shown by the black dots in each graph. Notice that in these figures, we show Kepler's third law (black lines) by first solving for the period: $P = \sqrt{A^3} = A^{1.5}$. The axes on the graph are logarithmic. On this kind of a graph, a relationship in which one variable is a power of another shows up as a straight line.

10. In order for Kepler's law to be considered a universal law, we must be able to apply it to every case where objects are orbiting a central body that is much more massive than they are. Based on Figure 5.5, summarize the evidence that Kepler's third law is most likely a universal law.

Step 5—Putting It Together

11. Describe how we use Kepler's third law when we are investigating the moons orbiting a planet, or planets orbiting stars other than the Sun. Be sure to use at least three key terms.

● ACTIVITY 6
Extraterrestrial Tourism

Learning Goals

In this activity, you will apply what you have learned about Kepler's laws and Newton's law of gravity to explore two systems in which planets orbit stars other than the Sun. After finishing this activity, you will be able to

1. compare two multiple-planet systems, Kepler-62 and Trappist-1, to the Solar System.

2. compare masses and radii of the planets of these two systems to those of Earth.

3. explain how orbital distances can be found from measurements of the orbital periods of exoplanets.

4. summarize how to estimate the mass of the planet.

5. combine the information for each planet in these systems, along with their equilibrium temperatures, to plan an imaginary space voyage.

Key terms: exoplanet, habitable, Kepler's third law, period, semimajor axis, astronomical unit (AU), Newton's law of gravity, weight, mass, equilibrium surface temperature

Step 1—Background

There is ample evidence that planetary systems—planets orbiting other stars—are common in the Milky Way galaxy. As of 2017, astronomers had discovered 2,657 planetary systems. Many of those systems contain multiple planets, so that the number of confirmed **exoplanets** was 3,504. There are more than 2,000 other "candidate" exoplanets; either the data for these have not yet been analyzed or the exoplanets need further observation. Of the confirmed exoplanets, 55 are considered potentially **habitable**. These exoplanets are more likely to have a rocky composition and conditions, such as temperature, favorable to maintain liquid water on the surface.

Imagine that we were able to travel to a potentially habitable planet. What information would give us confidence that we were making the right choice of where to go? In other words, how do we know what we know? For this activity, we need to assume that we have found a propulsion system fast enough to get the space travelers to either of the two systems we discuss in a "reasonable" amount of time. We find that we will need to travel at 0.999 times the speed of light.

Kepler's third law gives us the basic information. We measure the amount of time it takes a planet to orbit its star, its **period**. From that we can get the planet's average distance from the star (its **semimajor axis**). It's useful to express this in terms of the Earth's average distance from the Sun, the **astronomical unit (AU).**

1. Exoplanets that are more likely to have a rocky composition and conditions favorable to maintain liquid water on the surface are considered _____.

Step 2—An Overview of Two Multiple-Planet Systems

2. Study **Table 6.1**, which shows orbital and planetary data for the systems Kepler-62 and Trappist-1. State your overall conclusions about the planets' masses and radii (both compared to Earth's) for both planetary systems.

⬤ TABLE 6.1

Orbital and planetary data for the systems Kepler-62 and Trappist-1

PLANET	ORBITAL PERIOD (EARTH DAYS)	ORBITAL PERIOD (EARTH YEARS)	SEMIMAJOR AXIS (AU)	MASS (M_{EARTH})	RADIUS (R_{EARTH})
Kepler-62b	5.71	0.02	0.06	2.1	1.34
Kepler-62c	12.44	0.03	0.09	0.1	0.55
Kepler-62d	18.16	0.05	0.12	5.5	1.99
Kepler-62e	122.39	0.34	0.43	4.8	1.64
Kepler-62f	267.29	0.73	0.72	2.8	1.44
Trappist-1b	1.51	0.004	0.011	0.86	1.13
Trappist-1c	2.42	0.007	0.015	1.16	1.10
Trappist-1d	4.05	0.011	0.022	0.30	0.79
Trappist-1e	6.10	0.017	0.028	0.77	0.92
Trappist-1f	9.21	0.025	0.037	0.93	1.05
Trappist-1g	12.35	0.034	0.045	1.15	1.15
Trappist-1h	18.77	0.055	0.062	0.33	0.78

Data for Kepler-62 planets: http://www.openexoplanetcatalogue.com/planet/Kepler-62 b/ (accessed 10/08/18).
Data for Trappist-1 planets: http://www.trappist.one/#system (accessed 10/08/18).

3. In **Figure 6.1** the planets of the Kepler-62 and Trappist-1 systems are on the same scale as Mercury and Venus, as though they were part of the Solar System. Describe how the distances for these planets compare to the distances of planets from the Sun in our own system.

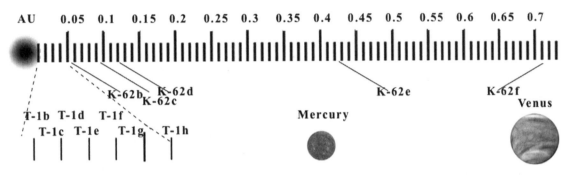

FIGURE 6.1

Step 3—Applying Kepler's Laws and Newton's Law of Universal Gravitation

In **Figure 6.2**, the orbital period is plotted versus the semimajor axis for both Trappist-1 (6.2a) and Kepler-62 (6.2b). The data for the systems are shown by the black dots. Notice that in these figures, we show Kepler's third law (black line) by first solving for the period: $P = \sqrt{A^3} = A^{1.5}$. The axes on these graphs are logarithmic. On this kind of graph, a relationship in which one variable is a power of another shows up as a straight line.

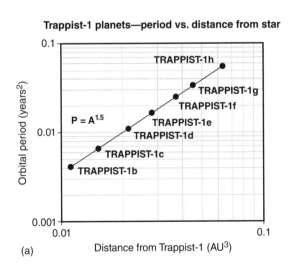

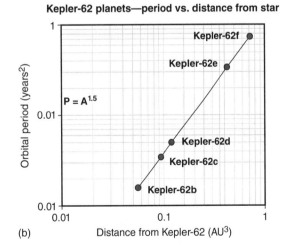

FIGURE 6.2

4. Kepler's third law originally described just the planets of the Solar System. How do the data for Kepler-62 support the universality of Kepler's third law?

5. Summarize how the orbital *distance* of a planet can be found from a measurement of the orbital *period* of the planet.

Now consider **Newton's law of gravity**, which describes how to calculate the force of gravity between two objects. The equal and opposite force between you and Earth is also known as your **weight**, and Newton's law of gravity gives:

$$\text{weight} = F_{Earth\&you} = G\frac{M_{Earth}}{R_{Earth}^2}m_{you},\text{ where } G \text{ is the gravitational constant.}$$

On any given planet or moon, your **mass** (m_{you}) will not change. It is something that depends only on how much "stuff" *you* have in your body. Your weight, however, will change depending on the mass, M, and radius, R, of the planet or moon you are on. To find your weight on another world, you would substitute its mass for the mass of Earth and its radius for the radius of Earth into Newton's law of gravity. But there is an easier way.

You can instead compare your weight on other worlds to your weight on Earth directly by taking ratios, to find that you will weigh twice as much (or a third as much) as you do on Earth. This mathematical "trick" divides two equations so that most things will cancel out, simplifying later calculations. Kepler-62b has 2.1 times the mass of Earth, so its mass can be written $2.1M_{Earth}$. Its radius is 1.34 times larger, so its radius can be written $1.34\,R_{Earth}$. To find how your weight on Kepler-62b would compare to your weight on Earth, first write Newton's law of gravity for each planet, and then divide the two equations. In practice, this means we divide the left side of the equals sign and the right side of the equals sign:

$$\frac{F_{K-62b}}{F_{Earth}} = \frac{G\dfrac{2.1M_{Earth}}{(1.34)^2 R_{Earth}^2}m_{you}}{G\dfrac{M_{Earth}}{R_{Earth}^2}m_{you}}$$

Just about every term on the right side of the equals sign $(G, M_{earth}, R_{earth}, m_{you})$ cancels. All that is left is:

$$\frac{F_{K-62b}}{F_{Earth}} = \frac{2.1}{(1.34)^2} = 1.2$$

You would weigh 1.2 times as much on Kepler-62b. We can use the same logic for the other planets, substituting the correct mass and radius for each. The results of these calculations are shown in **Table 6.2**.

○ TABLE 6.2

Mass, radius, and weight on selected extrasolar planets

PLANET	MASS (M_{EARTH})	RADIUS (R_{EARTH})	COMPARABLE WEIGHT
Kepler-62b	2.1	1.34	1.2
Kepler-62c	0.1	0.55	0.3
Kepler-62d	5.5	1.99	1.4
Kepler-62e	4.8	1.64	1.8
Kepler-62f	2.8	1.44	1.4
Trappist-1b	0.86	1.13	0.7
Trappist-1c	1.16	1.1	1.0
Trappist-1d	0.3	0.79	0.5
Trappist-1e	0.77	0.92	0.9
Trappist-1f	0.93	1.05	0.8
Trappist-1g	1.15	1.15	0.9
Trappist-1h	0.33	0.78	0.5

6. The most massive planet, Kepler-62d, is 55 times that of the least massive one, Kepler-62c. At first glance, this might make you think you would weigh 55 times as much on Kepler-62d as on Kepler-62c. However, if you calculate your weight, as in the example above, you would weigh six times as much on Kepler-62d as you would on Kepler-62c. Explain why you would not weigh 55 times more on Kepler-62c.

These planets were detected by the transit method, where the planet travels between its star and us and slightly dims the light we receive from the star. The radius of the planet can be estimated by how much light it blocks. To get the mass of the planet, we need to use Newton's version of Kepler's third law:

$$(M_1 + M_2) = \frac{4\pi^2}{G}\frac{A^3}{P^2}$$

where M_1 is the mass of the star, and M_2 is the mass of a planet. We get a good estimate of the mass of the star, M_1, by studying the star itself.

7. Explain how to determine the mass of the planet, M_2, once M_1 (the mass of the star) is known. Be sure to state what the symbols mean. Your explanation should be detailed enough to make it possible for another student in the class to understand how to do it.

Step 4—Planning Your Extraterrestrial Adventure

Tables 6.1 and 6.2 show that we now know a lot about these two planetary systems, and if you were to travel to either of these two stars, you would have a choice of which planet you would visit. You need a bit more information, though, in order to make a good decision.

Table 6.3 lists the **equilibrium surface temperatures** for the exoplanets in these systems that are considered habitable by astronomers. These temperatures are based on the exoplanet's distance from its star and an estimate of how much of its star's light it would absorb; nothing else—such as having an atmosphere—is assumed. Earth's equilibrium surface temperature is included in the table for comparison.

● TABLE 6.3

Equilibrium surface temperatures and related data for selected extrasolar planets

POTENTIALLY HABITABLE PLANET	EQUILIBRIUM SURFACE TEMPERATURE		STAR'S SURFACE TEMPERATURE	DISTANCE TO STAR FROM EARTH
	KELVIN	FAHRENHEIT		
Earth	255	−1	Sun: 5780 K	8.3 light-minutes
Kepler-62e	270	26	Kepler-62: 4925 K	990 light-years
Kepler-62f	208	−85	"	"
Trappist-1c	335	143	Trappist-1: 2550 K	40 light-years
Trappist-1d	282	48	"	"
Trappist-1e	246	−17	"	"
Trappist-1f	214	−74	"	"
Trappist-1g	195	−109	"	"
Trappist-1h	169	−155	"	"

8. After reviewing the information in Table 6.3, you should be able to compare the planetary systems and make your decision about which planet you will be traveling to. Identify that planet, list its characteristics, and state why you chose it. Explain why astronomers have considered it habitable based on the information given in this activity. (Recall that a planet is considered habitable if it is likely to have a rocky composition and conditions favorable to maintain liquid water on the surface. How does your chosen planet satisfy those requirements?)

9. Review Table 6.1. Will you weigh more or less on this planet than you do on Earth?

10. What will it be like living on your planet?

11. How will you get there? The fastest spacecraft made so far (the Helios spacecraft) traveled at 250,000 km/h (150,000 mph). At this speed, the Sun is about 600 hours (25 days) away. Kepler-62 is 37.5 billion hours (4.3 million years) away. Trappist-1 is 1.5 billion hours (170,000 years) away. Obviously, we need a faster spacecraft. Suppose that we invent a way to travel at 0.999 times the speed of light. At that speed (0.999 light-years per year), how long will it take (according to an observer on Earth) for you to travel to the stellar planetary system Kepler-62? (**Hint**: Kepler-62 is 990 light-years from Earth. You would divide that distance by 1 light-year per year if we could travel at light speed.)

12. How long would it take to travel to Trappist-1, traveling at 0.999 light-years per year?

Fortunately, according to Einstein's theory of special relativity, if you are moving in a spaceship traveling at a speed close to the speed of light, your time will pass more slowly. The details of special relativity will be saved for another time, but if the spaceships to Kepler-62 and Trappist-1 travel at a velocity of 0.999 the speed of light, you will age only 44 years during your travel to Kepler-62 and only 1.8 years to Trappist-1. It is a one-way trip, by the way.

Step 5—Putting It Together

The Jupiter-like planet orbiting the Sun-like star 51 Pegasi was discovered in 1995, and what a surprise this planet was. It is about half the mass of Jupiter but orbits much closer to its star than Mercury orbits the Sun: 0.05 AU versus 0.31 AU. At that time, we knew of two planetary systems: our Solar System and the single-planet star 51 Pegasi. The count now is more than 3,000, and we expect more exoplanets will be confirmed through ongoing analysis and the results of new surveys.

13. Review the steps in this activity and summarize how our knowledge of exoplanets has grown over the last few decades. Include in your summary how we have become confident in what we know about the exoplanets. Be sure to use at least three of the key terms in your explanation.

● ACTIVITY 7
Light and Spectra

Learning Goals

In this activity, you will explore the properties of continuous and line emission spectra. By working through this activity, you will be able to

1. explain how the temperature of an incandescent lightbulb affects the intensity and colors of the observed spectrum.

2. compare the observed continuous spectrum to a series of Planck (blackbody) curves.

3. examine emission spectra of five elements, noting the patterns and intensities of the lines.

4. identify three "unknown" elements.

5. state the significance of each element having its own unique spectral signature of emission lines.

Key terms: incandescent, luminous, blackbody, spectrum, continuous spectrum, wavelength, visible wavelength, infrared light, ultraviolet light, line emission, intensity

Step 1—Background

Incandescent bulbs pass an electric current through a resistive coil of wire, usually made of tungsten because of its high melting temperature. As the wire heats up, it starts to glow. The hotter the bulb is, the more **luminous** (brighter) it becomes. Incandescent radiation is thermal radiation (also called **blackbody** radiation). When an incandescent bulb is hot, it shines with the whole rainbow of colors. Splitting light into all of its separate colors creates a **spectrum**. These colors become brighter as the bulb gets hotter. Energy-inefficient incandescent bulbs are being phased out and replaced by the much more energy-efficient fluorescent and LED bulbs that do not depend on thermal radiation. The way incandescent bulbs shine mimics stars; fluorescent and LED bulbs do not shine in the same way.

When we see a rainbow in the sky, with all of the colors of the visible part of the spectrum—violet, blue, green, yellow, orange, red—we are looking at a **continuous spectrum**, one without a break in the colors. Light is traveling from the rainbow to us as waves of energy, waves characterized by **wavelength**, the distance from the peak of one wave to the peak of the next wave. The wavelengths we can see are known as **visible wavelengths**, and they range from about 350 nanometers (nm, 350 billionths of a meter) to about 750 nm. Light with wavelengths a bit longer than we can see is called **infrared light**. Light with wavelengths a bit shorter than we can see is called **ultraviolet light**.

If you were to look at the spectrum of a sodium-vapor street lamp (by shining its light through a prism or by reflecting its light carefully off a CD), you will see that only specific colors or wavelengths of light are present. The streetlight's spectrum has just a few curved, bright lines. We call those bright lines **line emission**. We could describe this by saying "the colors are more intense" at those wavelengths; the **intensity** of the light is greater.

Astronomers study the continuous and line emission spectra of stars to determine the temperatures and compositions of astronomical objects. In this activity, you will explore a few earthbound examples.

1. Rainbows show _____ spectra, while street lamps will have _____ spectra.

Step 2—Incandescent Bulbs and the Continuous Spectrum

Turn to **Figure 7.1** in the appendix. The image shows spectra (**Figure 7.1a**) from an incandescent bulb (**Figure 7.1b**), at its brightest at the top of the image and at its dimmest at the bottom of the image. These are continuous spectra because the colors run smoothly across the wavelengths. Use colored pencils—or work with shading with regular pencils—to reproduce the brightest spectrum in **Figure 7.2a** and the dimmest in **Figure 7.2b**. If you use a regular pencil, label each wavelength with the color that appears in the spectrum at that wavelength.

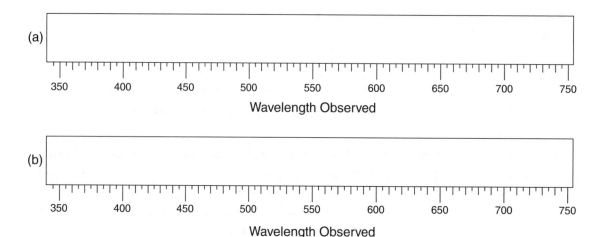

FIGURE 7.2

2. Is a dim incandescent bulb _____ **hotter** or _____ **cooler** than a bright incandescent bulb?

3. Write a sentence to describe how the temperature, as shown in **Figure 7.1c**, affects the bulb's intensity.

4. There are obviously colors missing in the spectrum when the bulb is at its dimmest compared with the spectrum of the bulb when it is at its brightest. Which colors are missing?

5. Write a sentence to describe how the temperature of the bulb affects the colors.

6. Combine your answers to questions 3 and 5 to write a sentence describing how the temperature affects BOTH the intensity and the colors in the spectrum of an incandescent bulb.

The more heat applied to the bulb, the more color appears as well as the stronger the light.

Step 3—Relating the Results to Blackbody Curves

The spectrum of light from a hot object can be graphed, with intensity (brightness) on the *y*-axis, and wavelength (related to color) on the *x*-axis. For objects that shine because they are hot, the resulting curve is called a "blackbody curve." A series of blackbody curves are shown in **Figure 7.3**, for an object at four different temperatures. These curves are "theoretical"; they were generated mathematically.

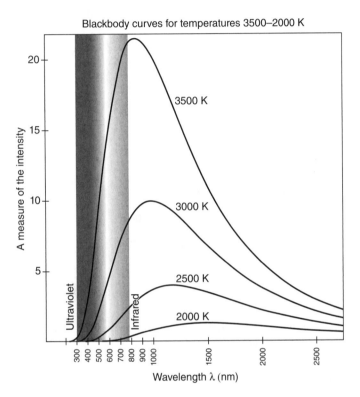

FIGURE 7.3

7. Do these theoretical curves support the summary you provided in question 6? Explain your answer, bringing in the intensity of the light at wavelengths in the visible part of the spectrum (300–700 nm).

8. The peaks in intensity for these curves occur in which region of the spectrum?

   a. ultraviolet   **b.** visible   **c.** infrared

The melting point for tungsten, the element for the filament of an incandescent bulb, is about 3700 K, which means that the filament probably does not get much hotter than about 3500 K. Examine **Figure 7.4**, which gives an expanded view of the blackbody curve for an object with a temperature of 2000 K.

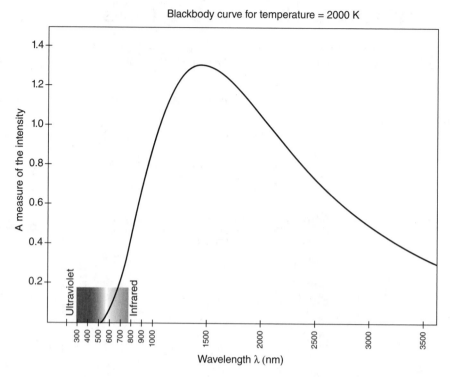

FIGURE 7.4

9. Assume that the temperature of the filament in the bulb at its dimmest was around 2000 K, so that its spectrum would look like the curve in Figure 7.4. Explain how we were still able to see some red in the spectrum of the incandescent bulb, even though the peak of the spectrum is well outside of the visible wavelength region.

10. If we could see at infrared wavelengths, would the bulb appear _____ **brighter** or _____ **dimmer**?

11. There are many stars that have surface temperatures in the 3500–2000 K range. Would they appear bright or dim to us? Explain your answer by contrasting a 3500 K star with a 2000 K star.

The 3500 K star would appear brighter compared to the 2000 K star, since 3500 is at the top end of visible wave lengths while 2000 is the faintest

Learning Astronomy by Doing Astronomy Second Edition

Step 4—Emission Spectra and Identifying "Unknown" Elements

Turn to **Figure 7.5** in the appendix. This series of images contains the true-color emission spectra of five elements. In **Figure 7.6**, use colored or regular pencils to reproduce these spectra. Be careful to note the wavelength at which each color appears and the intensity of each line. Indicate the intensity of a line in Figure 7.6 by making wider lines on your paper for brighter lines in the spectrum.

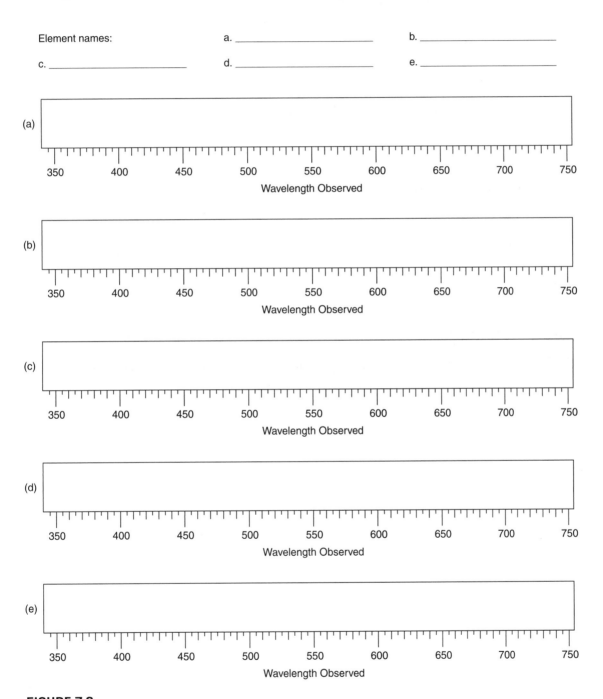

Element names: a. _____ b. _____

c. _____ d. _____ e. _____

FIGURE 7.6

12. Study the spectra that you sketched. Did any of the elements have the same emission spectrum? If so, which ones? If not, why not? *No, because each one is unique, an at least don't have the same properties with 3 cases*

13. Comment on both the similarities and differences you observed among these spectra. *Neon by far brightest others also seem more spread out.*

Turn to **Figure 7.7** in the appendix. This figure shows the lines of "unknown" elements. Consider the emission spectra you sketched in Figure 7.6 to be the reference spectra obtained in the laboratory. Identify each unknown element in the final series of images. The pattern of the lines, colors, and spacing are all important.

14. Based on your comparisons between the known and unknown elements, what are the mystery elements?

Step 5—Putting It Together

15. Describe how we can estimate the temperature of stars based on their blackbody (thermal) curve.

16. From your results in Step 4, summarize how spectra can be used to find the composition of a gas. Include in your summary the significance of each element having a unique spectrum. Be sure to include at least three key terms in your explanation. This method forms the basis for how astronomers determine the composition of stars—in other words, the stuff of which stars are made. *when looking at the spectrum of a gas your reading a unique pattern, this pattern put on top of other patterns will display each element present in the gas your observing at a wave lengths. Since each is unique emission and intensity is unique you can be sure of what you're seeing.*

● ACTIVITY 8
Deciding Where to Put the Telescope

Learning Goals

In this activity, you will learn about the impact of the atmosphere on astronomical observations and observatories. After completing this activity, you will be able to

1. describe how Earth's atmosphere affects decisions about which telescopes are built and where they are located.

2. distinguish among telescopes that can be used on the ground versus those that must be located in space in order to observe celestial objects at specific wavelengths.

3. explain what is meant by angular resolution and light-gathering power and why they are important.

4. summarize the advantages and disadvantages of ground-based telescopes and telescopes placed in orbit.

Key terms: electromagnetic spectrum, opaque, transparent, atmospheric window, observatory, space-based observatory, ground-based observatory, angular resolution, light-gathering power, aperture

Step 1—Background

We start our investigations about where to put a telescope (something that must come before the design of the telescope) based upon how Earth's atmosphere affects light passing through it. In some regions of the **electromagnetic spectrum** (light), Earth's atmosphere is **opaque**—it totally blocks all light at those wavelengths. In some regions, it is partially transparent—it lets some light through at specific wavelengths. In other regions of the spectrum, Earth's atmosphere is **transparent**—it lets all light at those wavelengths through. Transparent regions of the spectrum are called **atmospheric windows**. There are three windows identified in **Figure 8.1**: the visible window, infrared windows, and radio window.

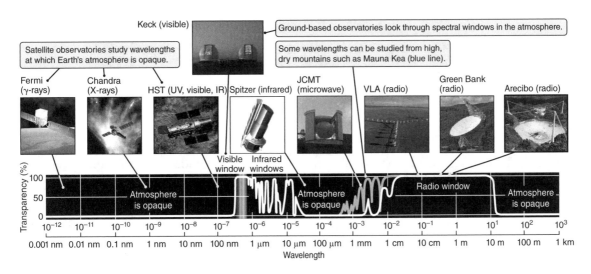

FIGURE 8.1

An **observatory** consists of more than just the telescope. Observatories include the structures that house the telescopes, the detectors that record the light, the optics that focus the light, and related instruments and equipment. Thus, the satellites (**space-based observatories**) shown in Figure 8.1 are considered observatories just like those situated on Earth (**ground-based observatories**). Above each observatory image is the name of the observatory plus the region of the spectrum it was designed to observe. From the shortest wavelength of light (highest energy) to the longest wavelength of light (lowest energy), these are gamma rays (γ-rays), X-rays, ultraviolet (UV), visible (what we can actually see), infrared (IR), microwave, and radio.

There are other considerations that must be addressed before a location is chosen. What are the celestial targets going to be? Do we need high **angular resolution** (measured as an angle of the celestial sphere) to see fine details of the objects? Are these objects extremely dim so that our telescope will need high **light-gathering power**? If so, we will require a large **aperture**, or light-collecting area. If we are monitoring the changing brightness of single stars in the Milky Way, then ground-based telescopes work fine. If we need high light-gathering power and angular resolution, then we might want to consider using a very large telescope located on a high mountain in a dry, dark place, surrounded either by desert or ocean for maintaining constant temperatures.

Examples of angular resolution can be seen in **Figure 8.2**. The resolution goes from (a) low—barely discerning what we are looking at; to better, as shown in (b); to (c), being able to possibly sketch and label the parts of this ancient brass telescope.

(a)   (b)   (c)

FIGURE 8.2

1. How does the atmosphere "being opaque" to a region of the electromagnetic spectrum differ from its "being transparent"?

Step 2—Earth's Atmosphere

2. Refer back to Figure 8.1. For which types of electromagnetic radiation is the atmosphere of the Earth completely opaque (transparency is 0 percent)?

---

3. Refer back to Figure 8.1. For which types of electromagnetic radiation is the atmosphere of the Earth completely transparent (transparency is 100 percent)?

4. Refer back to Figure 8.1. Identify regions of the spectrum that may be hard to classify as strictly opaque or transparent.

5. Astronomers often want to observe celestial objects in regions of the electromagnetic spectrum where the atmosphere is opaque. What options do they have?

Step 3—Telescopes

6. We now have telescopes to observe every part of the electromagnetic spectrum. Which type(s) of Earth-based telescopes can successfully observe celestial objects? (Select all that apply.)

   a. gamma-ray telescopes   **b.** X-ray telescopes   **c.** ultraviolet (UV) telescopes

   d. visible-light telescopes   **e.** infrared telescopes   **f.** radio telescopes

7. Which type(s) of telescopes *must* be placed and operated in space (are space-based) well above Earth's atmosphere? (Select all that apply.)

   a. gamma-ray telescopes   **b.** X-ray telescopes   **c.** ultraviolet (UV) telescopes

   d. visible-light telescopes   **e.** infrared telescopes   **f.** radio telescopes

Step 4—Angular Resolution and Light-Gathering Power

Let's consider the choices that need to be made between the advantages of getting the highest angular resolution and light-gathering power versus the disadvantages of the added cost, maintenance, repair, and other practical matters that will occur.

- The angular resolution of a telescope is directly proportional to ("proportional to" is given the math symbol $\propto$) the wavelength being observed and inversely proportional to the diameter of the aperture of the telescope:

$$\text{angular resolution} \propto \text{wavelength/diameter}$$

We want the angular resolution to be small (to detect very small details). This means we want to make the telescope as large as possible.

- The light-gathering power is proportional to the area of the aperture. Because the objective lens of a refracting telescope or the main mirror of a reflecting telescope is round, the light-gathering power is proportional to the radius squared. For example, if a telescope can be built with a diameter of 20 meters (radius of 10 meters), the light-gathering power compared to a 2-meter-diameter telescope (radius of 1 meter) is proportional to $\frac{10^2}{1^2} = 100$. The 20-meter telescope has 100 times more light-gathering power than the 2-meter telescope. That is a considerable increase.

8. When building a telescope, the goal is to achieve the smallest angular resolution and highest light-gathering power possible. There may be disadvantages to the telescopes required to meet those goals, such as cost or complexity. Discuss the pros and cons of the following types of telescopes and their locations. Consider both the wavelength region and the diameter when talking about getting the smallest angular resolution.

   a. A space telescope built to get the smallest angular resolution for radio wavelengths between 100 and 1,000 meters.

   b. An extremely large telescope with its main mirror 39 meters in diameter, built on a 3,046-m peak in Chile, to observe visual and near-infrared wavelengths.

Step 5—Putting It Together

9. Most of the highest-energy, most violent, calamitous events of the universe emit radiation at gamma-ray and X-ray regions of the spectrum. Building and launching space telescopes to put into orbit to observe these wavelengths costs billions of dollars. Based on the information in this activity, justify the expenditure.

10. There is a common misconception that we put astronomical telescopes into space or on top of the highest mountains possible because they will be much closer to the celestial objects they are observing. Using at least three of the key terms, explain the real reason why the observatories housing those telescopes are located in those places.

Name _____ Date _____ Section _____

● ACTIVITY 9
51 Pegasi—the Discovery of a New Planet

Learning Goals

In this activity, you will determine the orbital period and the radial velocity amplitude of a star due to the gravitational effects of its orbiting planet, and then use those data to find the radius of the exoplanet's orbit and a lower limit on its mass. After completing this activity, you will be able to

1. apply Kepler's and Newton's laws to find the orbital values and mass of an exoplanet.

2. compare the extrasolar planet to the planets of the Solar System.

3. test the current theory about the formation of the planet orbiting 51 Pegasi.

4. investigate whether a Jupiter-like gaseous planet can exist very close to its star.

5. state how we detect and interpret the motion of a star with one or more planets.

Key terms: exoplanet, radial velocity, wavelength, frequency, Doppler shift, spectroscopy, blueshift, redshift, semi-amplitude

Step 1—Background

In just the past few decades, astronomers have announced discoveries of 3,000-plus **exoplanets** orbiting stars in our galactic neighborhood. These discoveries should answer the question of whether the Solar System is unique. When astronomers state that they have discovered a new planet, they are saying that their data can best be interpreted as a planet orbiting a star. One cannot "prove" that these other planets exist; one can only state that, until the hypothesis is disproved, a planet orbiting the star best explains the observations. Using the **radial velocity** method of detecting exoplanets, astronomers measure the influence each one has on its parent star, the "wobble," as the star and planet orbit their common center of mass. Astronomers measure the changes in the **wavelengths** of the light from the star and translate them into changes in the radial velocity of the star—the motion directed toward or away from Earth.

What do we mean by the radial velocity method? Imagine how sound changes pitch as a siren on a police car approaches us and then speeds past. We hear a higher pitch (higher **frequency**; shorter wavelength) as the car approaches and a lower pitch (lower frequency; longer wavelength) as it moves away. Wavelengths are inversely proportional to frequencies, so if a higher frequency is observed, the light has a shorter wavelength, and vice versa. Like sound, light also travels in waves. We sketch the orbits of a planet and its star in **Figure 9.1**. The star is pulled in a small elliptical orbit by the gravity of the planet orbiting it.

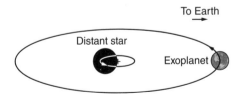

FIGURE 9.1

We show the orbit of the star in **Figure 9.2**. The wavelength observed, $\lambda_{\text{observed}}$, decreases as the star moves toward us (Figure 9.2b), and the wavelength, $\lambda_{\text{observed}}$, increases as the object moves away from us (Figure 9.2d). We call the shift in the wavelengths of light the **Doppler shift**. If the object is moving across our line of sight, it is neither going away nor coming toward us (Figures 9.2a and 9.2c), so the wavelength is not shifted—it equals the laboratory or rest wavelength, λ_{rest}.

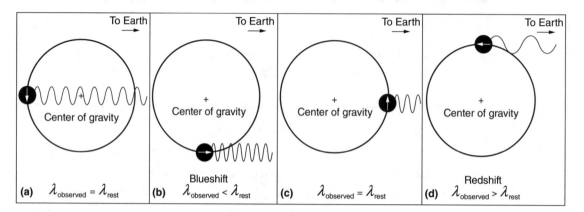

FIGURE 9.2

We measure the wavelengths of the star's light using **spectroscopy**, where the light is spread out into its component colors, and the spectral lines of the elements in its atmosphere are detected. If the star approaches, these spectral lines will have shortened wavelengths. The Doppler shift formula for solving for the velocity, v_r, is:

$$v_r = \frac{\lambda_{\text{observed}} - \lambda_{\text{rest}}}{\lambda_{\text{rest}}} \times c$$

Here v_r is the radial velocity of the object being observed and c is the speed of light, 3×10^8 m/s. The difference in the observed wavelength and the rest wavelength is *extremely small* for these planetary detections. If the observed wavelength is less than the rest wavelength, the radial velocity is negative, and we measure a **blueshift**. If the observed wavelength is greater than the rest wavelength, then the radial velocity is positive, and we measure a **redshift**. Here is an example for the red hydrogen line, one with $\lambda_{\text{observed}} > \lambda_{\text{rest}}$:

$$v_r = \frac{656.2805 \text{ nm} - 656.2801 \text{ nm}}{656.2801 \text{ nm}} \times 3 \times 10^8 \text{ m/s} = 183 \text{ m/s}$$

1. If the observed wavelength of the red hydrogen line were 656.2797 nm, has the line been redshifted or blueshifted? _____

Step 2—Analyzing the Observations

We enter this realm of discovery by working with actual data from observations of the star 51 Pegasi (51 Peg) made at the Lick Observatory in California. These data, shown as dots in **Figure 9.3**, come from the measurements of the Doppler shift of the absorption lines in the spectra of the star 51 Peg.

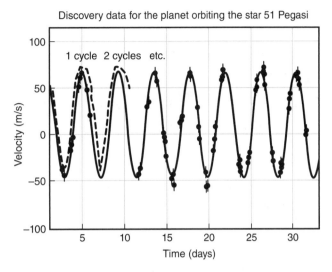

Discovery data for the planet orbiting the star 51 Pegasi

Examine the graph shown in Figure 9.3. It shows the measured radial velocities as a function of time recorded over a period of about 33 days. As you can see, the radial velocities are sometimes *positive* (the light is redshifted) and sometimes *negative* (the light is blueshifted), indicating that the star is moving away or approaching as viewed from Earth. This wobble of the star was the first indication that the star 51 Pegasi had an invisible companion, a planet officially known as Dimidium.

FIGURE 9.3

2. A period is defined as one complete cycle; that is, where the radial velocities return to the same position on the curve but at a later time. How many cycles did the star go through during the 33 or so days of observations?

    $\times$6   Number of cycles = _____7.5_____

3. What is the period, P, in days of one complete cycle? (Total number of days for these observations divided by number of cycles.)

    76   $P =$ ____3 5____ days   -49

4. What is P in years? (**Hint**: Divide the period in days by the number of days in a year; the answer will be a decimal number smaller than 1.)   52

    0.2 $P =$ ~~0.02~~ 0.01 years

5. What is the **semi-amplitude**, K? To find this, take 1/2 of the full range of the velocities. For example, we measure a maximum redshift velocity to be 25 m/s and a minimum blueshift velocity to be −29 m/s, making the full range 54 m/s. Divide that by 2 and we get K = 27 m/s.

    30   $K =$ ____50.5____ m/s

Because the mass of the star 51 Pegasi is close to that of the Sun, we can simplify the equations we use for determining the mass of Dimidium by calculating it in terms of Jupiter's mass and using the cube-root function on our calculators.

$$\frac{M_{\text{planet}}}{M_{\text{Jupiter}}} = \sqrt[3]{\frac{P}{12}} \times \left(\frac{K}{13}\right)$$

The period that you found, P, should be expressed as the decimal fraction of a year, and the semi-amplitude, K, that you found in m/s. Twelve years is the approximate orbital period for Jupiter and 13 m/s is the magnitude of the "wobble" of the Sun due to Jupiter's gravitational pull.

Learning Astronomy by Doing Astronomy Second Edition

How to solve: First divide the period you found as a fraction of a year by 12. That number becomes the "x" for the $\sqrt[3]{x}$ function you should find on your scientific calculator. Multiply that answer by the value of $\left(\dfrac{K}{13}\right)$. The answer we get for the entire equation is the ratio of the mass of the planet (M_{planet}) to the mass of Jupiter ($M_{Jupiter}$). As an example, let's let $P = 0.02$ yr and $K = 40$ m/s:

$$\frac{M_{planet}}{M_{Jupiter}} = \sqrt[3]{\frac{0.02}{12}} \times \left(\frac{40}{13}\right) = \sqrt[3]{0.00167} \times 3.08 = 0.12 \times 3.08 = 0.37$$

6. Plug in your values for P and K from questions 4 and 5, and calculate the mass of Dimidium in terms of the mass of Jupiter. Your calculations will give the mass of the planet as some number times the mass of Jupiter. For our example above: $M_{planet} = 0.37\ M_{Jupiter}$

$M_{planet} = \underline{0.46}\ M_{Jupiter}$

Because Dimidium is much less massive than its star, we can calculate the semimajor axis of its orbit in astronomical units (AU) using Kepler's third law:

$$P^2 = A^3$$

7. Expressing P as a fraction of a year, and A in AUs, solve for A:

$A = P^{2/3} = \underline{0.27}$ AU

How to solve: First square P. This number becomes the "x" for taking the cube root using the $\sqrt[3]{x}$ function on your scientific calculator.

Step 3—Comparing Results to Published Data

8. **a.** Compare your results with the set of published results as shown in **Table 9.1**. Make your comparisons quantitative by calculating the percent difference for each characteristic:

$$\frac{\text{your value} - \text{published value}}{\text{published value}} \times 100\%$$

○ TABLE 9.1

Comparison of your results with the published results for the planet

CHARACTERISTIC	PUBLISHED VALUE	MY VALUE	PERCENT DIFFERENCE
Mass (M)	0.47 $M_{Jupiter}$	0.46	2%
Orbital period (P)	4.23 days	5 days	18%
Semi-amplitude (K)	55.6 m/s	56.5 m/s	9%
Semimajor axis (A)	0.052 AU	~~0.27~~ 0.046	~~~~ 11%

b. Comment on this comparison. If your values are off by more than 50 percent, we suggest you review your measurements and calculations.

9. Compare the orbit of Dimidium around 51 Pegasi to the orbits of planets in the Solar System by referring to the rough scale model shown in **Figure 9.4**. Mercury is 0.4 AU from the Sun; Venus, 0.7 AU; Earth, 1.0 AU; Mars, 1.5 AU; Jupiter, about 5 AU. Where would Dimidium fit in *if* it were in our Solar System?

FIGURE 9.4

10. Science is based upon the ability to predict outcomes. However, nothing prepared astronomers for the characteristics of this "new" planetary system. Consider the mass of Dimidium as well as its distance from 51 Pegasi. Why was the discovery such a surprise when compared to our Solar System?

Step 4—Exploring the Observations

Because they were conscientious scientists, multiple teams of astronomers tried to disprove the hypothesis that the data indicated a planet was orbiting the star 51 Pegasi and suggested many alternate hypotheses. For example, this planet is very close to its central star, much closer than Mercury is to the Sun, and therefore will be extremely hot. Could this planet still be a gas giant at this distance from its star, or will it have lost its atmosphere? Because the planet is similar to Jupiter, and the star is similar to the Sun, we can simplify the calculations by imagining that Jupiter actually orbits the Sun at this much closer distance of 0.052 AU.

If we graph the temperature versus distance from the Sun in order to use Jupiter as our model, as shown in **Figure 9.5**, we can extrapolate to the temperature at 0.052 AU and apply that logic to Dimidium's orbit around 51 Pegasi. The temperature versus distance follows the inverse-square law.

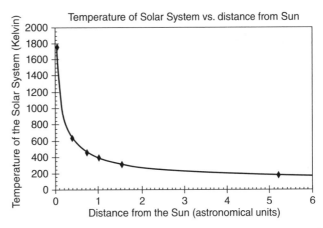

FIGURE 9.5

From this graph, we find that the temperature of Dimidium is 1800 K. We need to apply this to Jupiter so we can find out if this temperature would destroy the hydrogen atmosphere of Jupiter. This involves determining the velocity of the molecules of hydrogen gas in its atmosphere.

11. Using the molecular mass of hydrogen, 2, and the temperature we determined, calculate the velocity of the hydrogen gas molecules:

$$v_{gas}(m/s) = 157\sqrt{\frac{\text{Temperature}}{\text{Molecular mass}}} = 157\sqrt{\frac{1800}{2}} = \text{_____ } m/s$$

The escape velocity, $v_{esc\,Jupiter}$, for Jupiter is about 60,000 m/s. The condition for a planet to hold onto its atmosphere for billions of years is that the gas velocity, v_{gas}, must be less than one-sixth of the escape velocity for the planet: $v_{gas} < \frac{1}{6}v_{esc}$.

12. Is this condition satisfied? That is, is the gas velocity less than 1/6 the escape velocity? (In working with Jupiter, is $v_{gas} < 10{,}000$ m/s?) _____

13. Given that the 51 Peg planetary system is similar to ours, but with just one Jupiter-like planet, if Dimidium formed at a distance far enough away to originally have a substantial molecular hydrogen atmosphere, would it retain that atmosphere as it migrated to its current distance of 0.052 AU? How do your results support the conclusion that its planet is probably gaseous, like Jupiter?

Step 5—Putting It Together

14. Using the key terms for this activity, explain how astronomers determine the radial velocity of a star and how they know when the star is coming toward us and when it is going away from us. How do these measurements of any given star lead to the conclusion that there is a planet orbiting that star?

15. There are at least 500 confirmed exoplanets whose masses are greater than Jupiter's mass and that are located closer than 0.1 AU from their star. (Recall Mercury is 0.4 AU from the Sun.) Observational evidence strongly implies that these planets are gaseous, like Jupiter. In their planetary systems, where did these planets likely form and how did they get to be so close to their stars?

● ACTIVITY 10
Ranking the Steps of Planet Formation

Learning Goals

In this activity, after ranking the events that take place during star and planet formation, you will be able to

1. correctly order the stages of the formation of the Solar System.

2. identify the stages where overlapping of or simultaneous events occur.

3. recount the modern theory of planetary system formation.

Key terms: ecliptic, axial tilt, nebular theory, molecular cloud, self-gravity, gravitational potential energy, conservation of angular momentum, protoplanetary, planetesimal

Step 1—Background

We start this activity with a list of observations of the contents and characteristics of the Solar System, observations that must be explained by any theory of how this planetary system formed.

- Planets are located on or close to the **ecliptic**, the plane of the Earth's orbit around the Sun.

- The ecliptic plane is closely aligned with the equator of the Sun.

- Planets all orbit in a counterclockwise direction, as observed from the Northern Hemisphere perspective, and this is the same direction as the Sun rotates.

- Most of the planets rotate in a counterclockwise direction as well, but not all.

- The rotation **axial tilts** (the angle between the rotation axis and the plane of the orbit) for Venus and Uranus are large with respect to their orbital planes.

- There are distinct differences between the inner and outer planets in size and composition.

- The planets and moons with solid surfaces all have evidence of cratering.

- There is an asteroid belt where a planet should be.

- Meteorites show evidence of being made up of even smaller chunks of material.

While it may seem that our planetary system is just way too complicated to ever have a simple logical theory of how it formed, the **nebular theory**, which describes how the Solar System formed from a cloud of dust and gas, does just that.

1. Pick two of the above observations that could be observed from Earth without any special equipment.

Step 2—Visualizing the Formation of the Solar System

It helps to visualize the process involved in the nebu-lar theory. We start with the birth of a star and planets within a giant, cool, rotating **molecular cloud** of dust and gas operating under its own **self-gravity** (**Figure 10.1**). **Gravitational potential energy** is transformed during the contraction into heat.

Conservation of angular momentum states that if a spinning object shrinks, then the velocity of its spin must increase. The progression is seen in **Figure 10.2**, with the last image showing the emerg-ing Sun and planetary disk at the center. The word "proto" basically means "this comes first and then develops into something else," giving the "baby" Sun

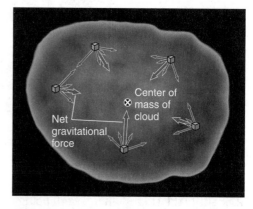

FIGURE 10.1

the name "protostellar" Sun that formed in the center of the "**protoplanetary**" disk (**Figure 10.3**). This disk started out with small particles of gas and dust. Eventually, "**planetesimals**" (tiny planets) formed (**Figure 10.4a** and **b**). Recall that the more mass an object has, the more gravity it exerts on other objects. Collisions happen. The protostellar Sun becomes the hottest object in the system, with temperatures dropping quickly farther away in the protoplanetary disk (**Figure 10.5**).

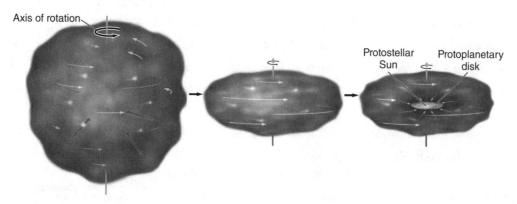

FIGURE 10.2

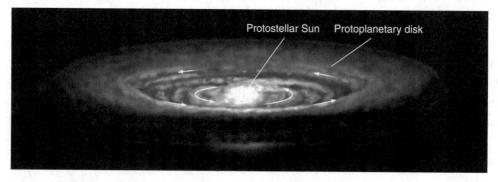

FIGURE 10.3

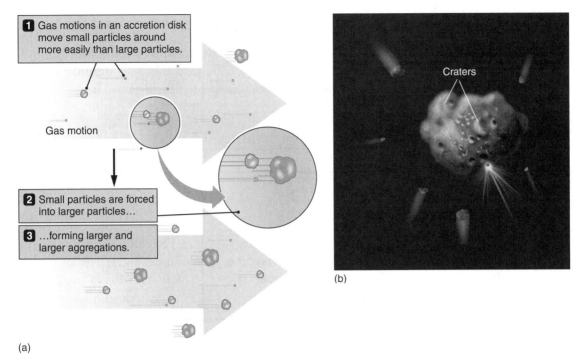

1 Gas motions in an accretion disk move small particles around more easily than large particles.

Gas motion

2 Small particles are forced into larger particles…

3 …forming larger and larger aggregations.

Craters

(a)

(b)

FIGURE 10.4

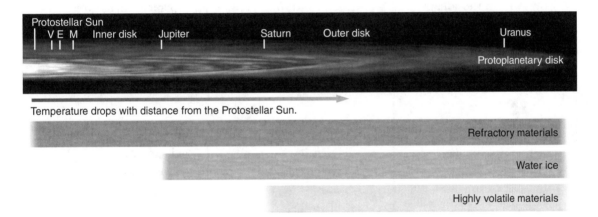

Protostellar Sun
V E M   Inner disk   Jupiter   Saturn   Outer disk   Uranus

Protoplanetary disk

Temperature drops with distance from the Protostellar Sun.

Refractory materials

Water ice

Highly volatile materials

FIGURE 10.5

Step 3—Ranking Steps in the Formation of a Planetary System

2. Arrange these steps in the order that they would occur according to the nebular theory of planet formation. If events can happen simultaneously, rank them equally in the order. Steps A and I are in the correct order as the first and last steps of those shown here. Write the ranking here:

A _____ _____ _____ _____ _____ _____ _____ I

A	B	C
A cloud of interstellar gas and dust starts to collapse under the force of its own self-gravity.	Motions push grains back and forth past other grains; smaller grains stick to the larger grains.	The cloud of gas rotates faster and faster due to conservation of angular momentum.
D	**E**	**F**
Gravitational potential energy from the collapse is converted into heat.	Material makes its final inward plunge, landing on a spinning, protoplanetary disk.	Accretion continues, with planetesimals growing in mass and becoming planets.
G	**H**	**I**
Planetesimals form that have enough gravity to attract other planetesimals.	Inner parts of cloud begin to fall freely inward, raining down on the center.	Inner planets become small and rocky while outer planets form gaseous giants.

Step 4—Putting It Together

3. Based on the ranking of the stages, write a summarizing paragraph that "tells the story" of how the Solar System formed. Be sure to use as many of the key terms as possible.

◉ ACTIVITY 11
Finding Surface Ages from Crater Counts

Learning Goals

In this activity, you will find the relative ages of regions on the Moon by comparing the distribution of crater sizes in different regions. You will be able to do this by

1. summarizing the general surface features of the Moon.

2. setting criteria for measuring the sizes of craters on the Moon.

3. determining the distribution of sizes for both heavily and lightly cratered regions.

4. stating the relative ages of the heavily and lightly cratered regions of the Moon.

5. evaluating the effectiveness of this method for dating cratering events in the Solar System.

Key terms: mare, crater, relative age, number density, resurfacing, hypothesis

Step 1—Background

Perhaps we take the Moon for granted when we look for celestial objects to study. It takes only a low-power telescope or binoculars to see details of its **mare** (relatively smooth "seas" of basalt) and **craters** large and small that indicate impacts from asteroids and meteoroids starting billions of years ago.

Planetary scientists count the number of craters found in defined areas of planets and moons in order to estimate **relative ages** of those areas. These counts are expressed as a **number density**. The number density differs from the more common use of the word "density" that means mass divided by volume. Number density for this activity means: "The number of craters of different sizes that are found within an area of a particular size." Thus, we can talk about "high" and "low" number density.

We assume that there would be no select region of any planet or moon that would experience a greater impact rate than any other region—all areas would be equally exposed to rocks from space. Planetary scientists talk about "resetting the clock" when **resurfacing** occurs. For example, while the Earth itself is about 4.6 billion years old, the land being created by lava flowing from Mauna Loa in Hawaii is "just being born." Geologically, the age of this new land can be just a few minutes old.

We will closely study two contrasting regions of the Moon, and then extend our results to make a general statement about the Moon. As scientists we need to understand the fundamentals of any method before we have enough confidence to use it.

1. Examine the upper-left region of the mosaic of images of the Moon shown in **Figure 11.1** to that of the lower-right region. How do the number densities of craters in each area compare?

2. What do your observations imply about the relative ages of those two regions?

3. Combine your answers to questions 1 and 2 and write your **hypothesis** by finishing this sentence: "If the number density in the upper-left region of the image has a lower number density of craters than the lower-right, then . . ."

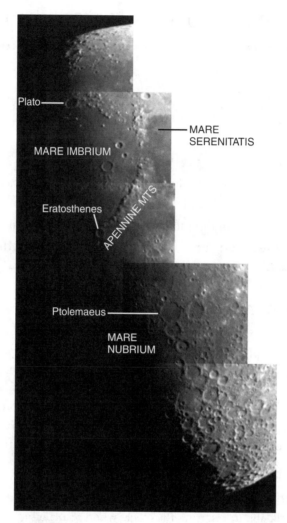

FIGURE 11.1

Step 2—The Cratered Highlands

Examine the centimeter scale and circle sizes included in **Figure 11.2**. Locate the reference craters in **Figure 11.3**: Albategnius (136 km diameter), Arzachel (97 km), and Alpetragius (40 km).

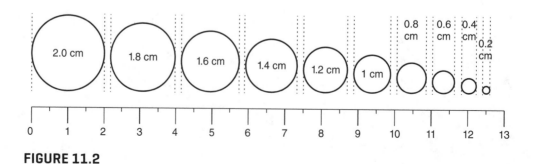

FIGURE 11.2

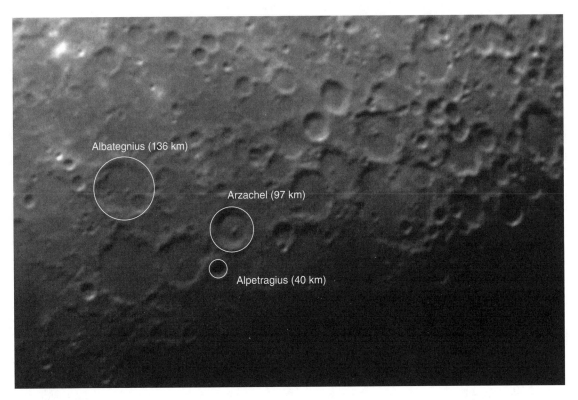

FIGURE 11.3

4. The image scale describes how many centimeters on the image correspond to a kilometer in space. Calculate the image scale by dividing the size of each crater in kilometers by its size in the image in centimeters. For example, consider a crater known to be 150 km in diameter that you measure to be 1.5 cm on the image. The image scale factor for this example is

$$1.5 \text{ cm} = 150 \text{ km}, \frac{1.5 \text{ cm}}{1.5} = \frac{150 \text{ km}}{1.5}, \text{giving us an image scale factor of } 1 \text{ cm} = 100 \text{ km}.$$

5. For each of the crater sizes listed in the "Cratered Highlands" section of **Table 11.1**, enter the corresponding diameter in kilometers by applying the average scale factor.

6. In Figure 11.3, count the number of craters of each size. Insert the results in Table 11.1. Be careful not to count craters twice. Count even the faintest craters you can see and those that overlap other craters. Because you are restricting your measurements to what are known as "binned" sizes, pick the closest size as you enter the counts in the table. If you do not locate a crater of a given diameter, enter 1 for number counted. As with all measurements, it is important that you are consistent in the methods you use.

Calculate the image scale factor for each of the three reference craters and average the values. Enter the average value for the scale factor into the "Cratered Highlands" section of Table 11.1.

⊙ TABLE 11.1

Crater counts for two regions of the Moon

CRATERED HIGHLANDS			UPPER MARE REGION		
IMAGE SCALE FACTOR: 1 CM = _____ KM			IMAGE SCALE FACTOR: 1 CM = _____ KM		
MEASURED SIZE (CM)	ACTUAL SIZE (KM)	NUMBER COUNTED	MEASURED SIZE (CM)	ACTUAL SIZE (KM)	NUMBER COUNTED
2 and >2			2 and >2		
1.8			1.8		
1.6			1.6		
1.4			1.4		
1.2			1.2		
1.0			1.0		
0.8			0.8		
0.6			0.6		
0.4			0.4		
0.2 and < 0.2			0.2 and < 0.2		

Step 3—The Upper Mare Region

7. Following the listed steps above, fill out the right-hand side of Table 11.1 by measuring craters seen in **Figure 11.4**, which shows a small part of the upper mare region of the Moon. You will need to find the image scale for this image using the three reference craters, Theaetetus (25 km diameter), Aristillus (55 km), and Archimedes (83 km). (The image scale factors should be the same, but you need to check.)

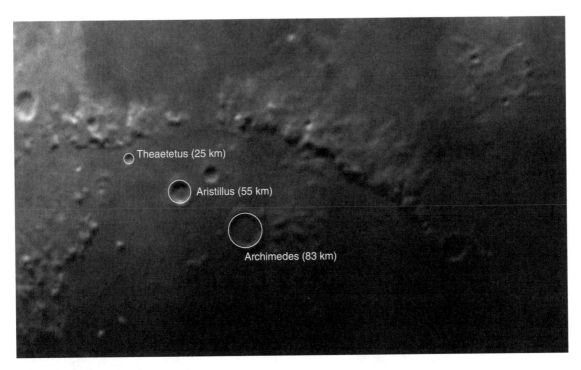

FIGURE 11.4

Step 4—Graphing the Data and Drawing Conclusions

The graph shown in **Figure 11.5** uses a logarithmic scale for the *y*-axis and a linear scale for the *x*-axis. The diagonal lines show the surface ages based on crater counts, from 3.9 billion years (giga-years; Gyr) down to 3.7 billion years. Scientists used radiometric dating of Moon rocks brought back by *Apollo* astronauts to find these actual ages. The sunlit part of the images you measured cover just about 1 million square kilometers, so you can compare your results to published results, as the diagonal lines represent counts versus diameter over 1 million square kilometers as well.

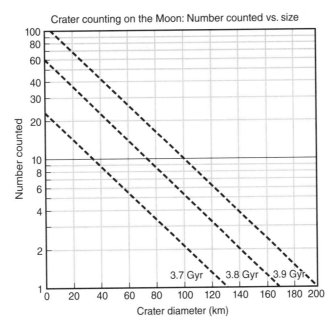

FIGURE 11.5

8. Graph the number of craters you counted versus actual size for both the cratered highlands region and the upper mare region. Use different symbols for the two regions. Make a note of the symbols you used here:

    Symbol used for cratered highlands: _____

    Symbol used for upper mare region: _____

9. What do the numbers and densities of the craters imply about the relative ages of these two regions of the Moon? Do your results support the hypothesis you made in Step 1? Comment.

10. Comparing crater densities on a terrestrial world easily leads to statements of *relative* ages, but relating crater densities to *actual* ages for the Moon required our sending astronauts there to bring back rocks to be dated based on radioactive isotopes. This is how the straight lines in Figure 11.5 are determined. Based on your results, what are the approximate ages or age ranges for the cratered highlands region and the upper mare region?

Step 5—Putting It Together

11. Evaluate this method for finding the relative ages and for estimating the actual ages of the surfaces of the Moon. What are the advantages and disadvantages in finding relative versus actual ages?

12. Using at least three of the key terms, summarize the method whereby we are able to determine relative ages of regions of the Moon based on crater-counting statistics.

● ACTIVITY 12
Planning a Manned Mission to Mars

Learning Goals

In this activity, you will be using orbital data and knowledge of Kepler's third law to formulate a plan for a manned mission to Mars with a successful return to Earth. After completing this activity, you will be able to

1. find and interpret trends in the dates of Mars oppositions.

2. formulate relationships between opposition dates and launches of spacecraft.

3. use Kepler's third law to understand the concept of a Hohmann transfer orbit.

4. apply Kepler's laws to the possibility of a manned flyby mission to Mars.

5. determine planetary positions for a launch from Mars to return successfully to Earth.

6. summarize the practical considerations for a manned round-trip mission to Mars.

Key terms: flyby, orbiter, lander, rover, weightless, Hohmann transfer orbit, eccentricity, perihelion, astronomical unit (AU), aphelion, opposition

Step 1—Background

The first missions to Mars were launched in the early 1960s by the United States and by the then Soviet Union. During the early years, there were more failures and partial successes than successes by both nations. Since that time, there have been a number of **flybys** (the spacecraft kept going past Mars), **orbiters** (the spacecraft went into orbit around Mars), **landers** (the spacecraft arrived on the surface of Mars), and **rovers** (the spacecraft had a component that moved around on the surface of Mars). Currently, there are about nine landers, rovers, and orbiters at Mars. Private enterprises and NASA are considering plans for a manned mission.

Current estimates for sending a privately funded manned mission to Mars are in the $1–$3 billion range per astronaut. The logistics of how to keep astronauts healthy under **weightless** conditions for long periods of time (they would be "freefalling" to Mars) and then how to support a team living on Mars are mind-boggling.

In this activity, we look at just one aspect of a manned mission to Mars: how to get the astronauts there and how to get them back using a **Hohmann transfer orbit**. A Hohmann transfer orbit means a spacecraft from a planet with a circular orbit (**eccentricity** equals 0) journeys to another planet with a circular orbit in the most energy efficient way.

Kepler's third law states that the period of an orbit squared is equal to the semimajor axis of that orbit cubed: $P^2 = A^3$. This is true for a manned spaceship orbiting the Sun. A spaceship traveling from Earth to Mars will have its **perihelion** (when it is closest to the Sun) at 1 **astronomical unit (AU)** and its **aphelion** (when it is farthest from the Sun) at 1.52 AU, Mars's orbital distance.

We start by looking at characteristics of the orbit of Mars relative to that of Earth, and what the history of past launches might tell us about planning a future launch.

1. Why is stating that the astronauts would be "freefalling" to Mars more accurate than saying they are in zero gravity?

Step 2—Relating Launch Dates and Opposition Dates

When scientists work with data, they look for trends and relationships among the variables. The graph shown in **Figure 12.1** plots the approximate dates of the **oppositions** of Mars. Mars is in opposition when the alignment is Mars-Earth-Sun; that is, Mars is opposite the Earth from the Sun.

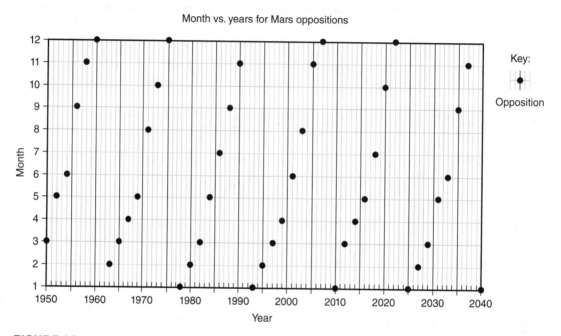

FIGURE 12.1

2. There are definite trends in these data. Give your interpretation of the trends in the opposition dates for Mars. (**Hint:** Check how much time passes between oppositions.)

There also are trends in the launch dates that appear to be related to the opposition dates of Mars. The dates of the launches of spaceships from Earth (for a few oppositions there were multiple launches) have been added to the opposition dates from Figure 12.1 and shown in **Figure 12.2** as squares. The squares marked with triangles indicate launches for the Mars opposition the following year. The launch data represent launches from many nations, not just the United States.

Dashed lines for the oppositions occurring in 1969, 1973, 1988, and 1999 show that launches were made between one and three months prior to that year's opposition.

3. There were 13 launches that occurred in the same calendar year as an opposition, and another seven launches for oppositions occurring the next calendar year (marked with triangles). Find two oppositions that happened after the year 2000 that had launches in one or more months preceding it, connect the launches with the Mars opposition as we have shown with dashed lines, and fill in **Table 12.1** with your data.

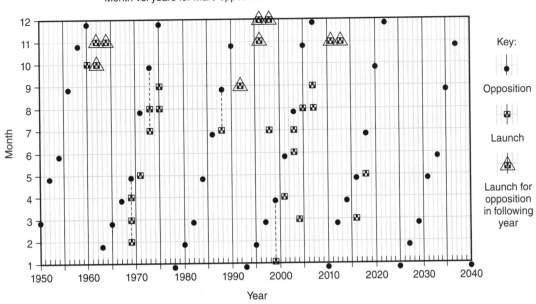

FIGURE 12.2

○ TABLE 12.1

Comparisons of oppositions of Mars and launches from Earth

OPPOSITION YEAR	NUMBER OF LAUNCHES	SUMMARY OF LAUNCHES PRIOR TO OPPOSITION
1969	3	3, 2, and 1 month before
1973	2	3 and 2 months before
1988	1	2 months before
1999	1	3 months before

4. Review the data in Table 12.1 as well as other oppositions and launches. What is the relationship between an opposition and any launches that occurred just prior to it? Why does that relationship exist?

Step 3—Developing a Simplified Model to Get to Mars

Kepler's laws can be used to estimate the year and month when there will be future "windows" for a launch. This is why Figures 12.1 and 12.2 show oppositions up to the year 2040.

A Hohmann transfer orbit is the most efficient way to transfer a spaceship from Earth to Mars and back again. A simplified transfer orbit is shown in **Figure 12.3**.

Learning Astronomy by Doing Astronomy Second Edition

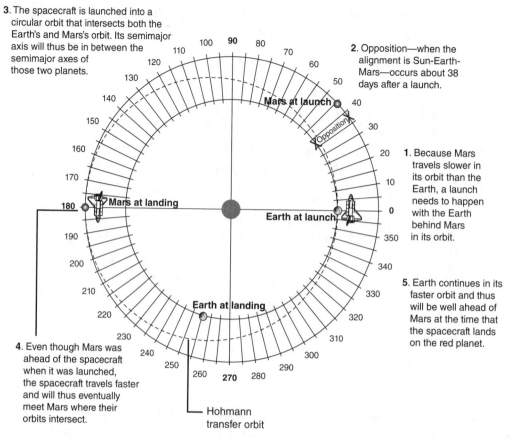

3. The spacecraft is launched into a circular orbit that intersects both the Earth's and Mars's orbit. Its semimajor axis will thus be in between the semimajor axes of those two planets.

2. Opposition—when the alignment is Sun-Earth-Mars—occurs about 38 days after a launch.

1. Because Mars travels slower in its orbit than the Earth, a launch needs to happen with the Earth behind Mars in its orbit.

5. Earth continues in its faster orbit and thus will be well ahead of Mars at the time that the spacecraft lands on the red planet.

4. Even though Mars was ahead of the spacecraft when it was launched, the spacecraft travels faster and will thus eventually meet Mars where their orbits intersect.

Hohmann transfer orbit

FIGURE 12.3

5. Explain the relationship between the dates of launches and Mars's opposition dates based on the information given in Figure 12.3.

There is a troubling fact about sending humans to Mars—it may need to be a one-way trip. Figure 12.3 shows that Earth is well beyond Mars when the spaceship lands on Mars. If the spaceship flies by Mars, and keeps going, will it meet Earth again on the Hohmann transfer orbit?

6. We opted for a flyby mission to Mars instead of a landing. Use the data given in **Table 12.2** to figure out where Earth and Mars will be after 518 days have gone by—the orbital period for the spaceship. Mark those locations on **Figure 12.4**. Sample calculations are given on the facing page. Show your calculations on this page, as well as the locations of Mars and Earth.

● TABLE 12.2

Orbital characteristics of the Earth, a spaceship to Mars, and Mars

OBJECT	SEMIMAJOR AXIS (A, AU)	PERIOD (P, YEARS)	PERIOD (DAYS)	DEGREES TRAVELED PER DAY
Earth	1	1	365.25	0.986
Spaceship	1.26	1.42	518	0.695
Mars	1.52	1.88	687	0.524

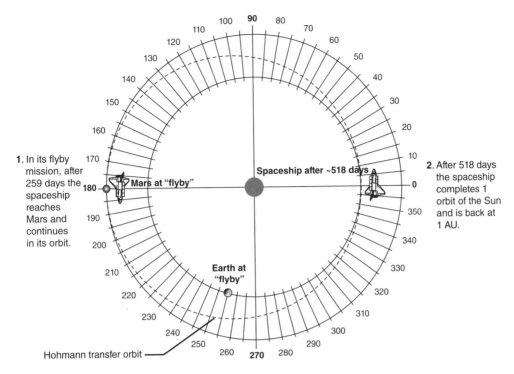

1. In its flyby mission, after 259 days the spaceship reaches Mars and continues in its orbit.

Mars at "flyby"

Spaceship after ~518 days

2. After 518 days the spaceship completes 1 orbit of the Sun and is back at 1 AU.

Earth at "flyby"

Hohmann transfer orbit

FIGURE 12.4

Sample calculation: Suppose an asteroid is orbiting between Earth and Mars that has a semimajor axis of 1.2 AU. Under Kepler's third law, this means an orbital period of 1.3 years, or about 480 days. The degrees traveled per day would be $\dfrac{360°}{480 \text{ days}} = 0.75°$ per day. After 518 days, the asteroid would have traveled 518 days $\times \dfrac{0.75°}{\text{days}} = 388°$. It would have gone around more than once.

7. Summarize why the astronauts in the spaceship are not happy, and may be panicking, when they return to 1 AU, to the launch location in their orbit.

Step 4—Planning the Return of the Manned Mission

How long will we need to stay on Mars until the planets are in position for a Hohmann transfer orbit back to Earth? Where will Earth need to be in its orbit when we launch from Mars? Once again, we use the results of Kepler's third law as listed in Table 12.2 and put our answer into **Figure 12.5**. Questions 8–11 on the next page help you figure this out. Answer those first, then place Earth on this figure.

The spaceship will take about 259 days to return to Earth's orbit from Mars. Earth travels 0.986 degrees per day. Where will Earth need to be in its orbit at the time of our launch from Mars so that the spaceship meets Earth in its orbit 259 days later?

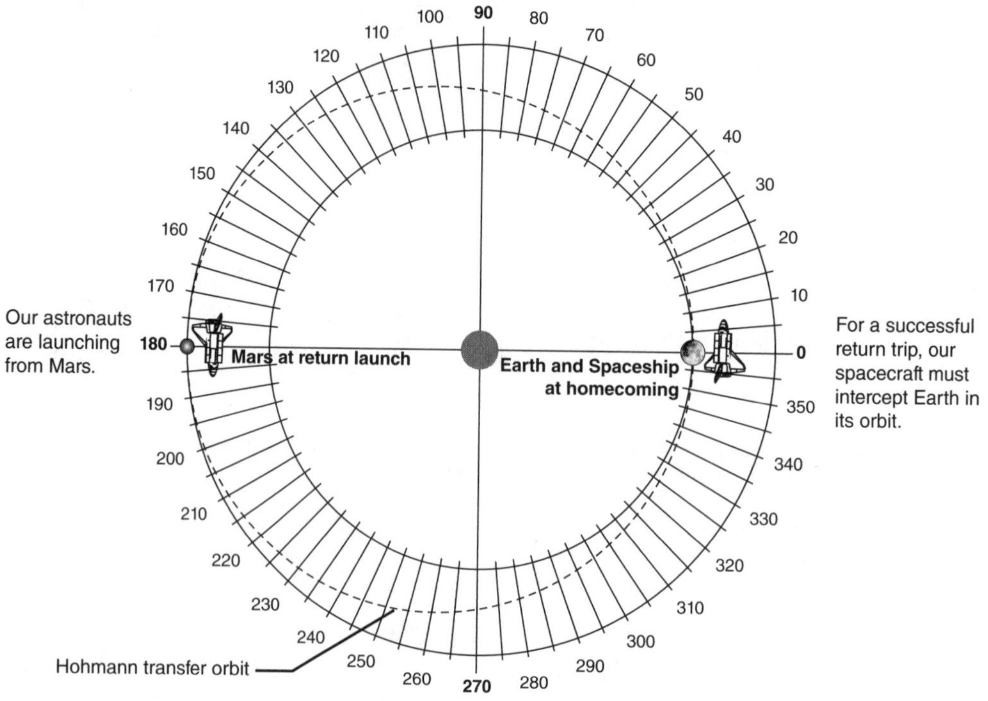

Our astronauts are launching from Mars.

Mars at return launch

Earth and Spaceship at homecoming

For a successful return trip, our spacecraft must intercept Earth in its orbit.

Hohmann transfer orbit

FIGURE 12.5

8. Find the number of degrees Earth travels in 259 days:

$$259 \text{ days} \times 0.986 \text{ degrees per day} = \underline{\hspace{2cm}} \text{ degrees}$$

9. This is how far Earth will need to travel to reach 0 degrees. When the spaceship launches from Mars, Earth should be at 360 minus this number of degrees, which is equal to _____ degrees. Mark this location on Figure 12.5.

10. Mars moves 0.524 degrees per day. How far will Mars travel while the spaceship is traveling to Earth?

$$259 \text{ days} \times 0.524 \text{ degrees per day} = \underline{\hspace{2cm}} \text{ degrees}$$

11. Where will Mars be when the spaceship arrives at Earth? _____ Mark this location on Figure 12.5.

12. Will the astronauts be able to look up and see Mars in the sky once they get home, or will Mars be opposite the Sun? Explain.

Step 5—Putting It Together

13. Using some of the key terms and your answers to the questions in this activity, summarize the practical limitations of using a Hohmann transfer orbit for a successful launch to and return from Mars of a manned mission.

● ACTIVITY 13
Planetary Climates

Learning Goals

One of the most useful skills that you can take away from any science class is to become proficient at reading graphs. Right now, one of the most important topics that you can learn about is the evidence of climate change. In this activity, you put the two together, using graphs of climate variables to learn about both graphs and climate change. After completing the activity, you will be able to

1. explain how we know that atmospheres affect the temperature of planets.
2. explain why it is the number of greenhouse gas molecules (rather than the percentage) that is important to know when considering planetary climates.
3. summarize why and how greenhouse gases affect the temperature of a planet.

Key terms: thermal equilibrium, greenhouse gas, opaque, carbon dioxide, greenhouse effect, concentration

Step 1—Background

In **thermal equilibrium**, the amount of energy that leaves a planet exactly balances the amount that is absorbed by it, as shown in **Figure 13.1**. If the planet receives "extra" energy, it will heat up, and therefore radiate more energy, until it comes back into thermal equilibrium at a new temperature. If a planet receives less energy, it will cool down, and therefore radiate less energy, until it comes back into thermal equilibrium at a new temperature. Thermal equilibrium is the fundamental physical principle that determines the surface temperature of a planet.

Greenhouse gases are gases that absorb in the infrared region of the spectrum. You may already know that infrared telescopes are placed in space because Earth's atmosphere is **opaque** in the infrared region of the spectrum; that is, most infrared radiation is blocked by the atmo-

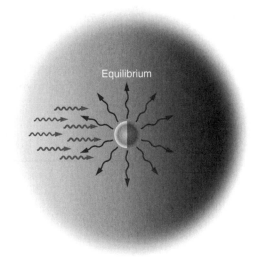

FIGURE 13.1

sphere. From these two facts, you might reason that Earth's atmosphere contains greenhouse gases, and you would be right. These gases absorb infrared radiation entering Earth's atmosphere from space, and likewise, they absorb infrared radiation trying to leave Earth's surface from the ground.

This happens on other planets as well. Venus and Mars have nearly identical compositions of their atmospheres: more than 95 percent **carbon dioxide**, which is a greenhouse gas. Yet Venus and Mars have very different temperatures. In this activity, you will figure out why that is so, and consider the implications for Earth.

1. If energy is delayed from leaving a planet, how will the planet's temperature change?

Step 2—Equilibrium Temperatures of the Planets

If we set the energy received from the Sun equal to the energy emitted from the planet, we can solve for how warm the planet should be, based solely on the energy that it receives from the Sun. These temperatures are given in **Table 13.1**, along with the minimum and maximum observed global temperatures.

⊙ TABLE 13.1

Equilibrium temperatures of the terrestrial planets

PLANET	DISTANCE FROM SUN (AU)	CALCULATED EQUILIBRIUM TEMPERATURE (K)	MINIMUM OBSERVED TEMPERATURE (K)	MAXIMUM OBSERVED TEMPERATURE (K)
Mercury	0.387	434	100	725
Venus	0.723	232	733	737
Earth	1	255	184	292
Mars	1.524	218	133	293

2. Graph the calculated equilibrium temperature from Table 13.1 on the axes shown in **Figure 13.2**. Use a filled-in circle to mark the spot on the graph, and label the marks.

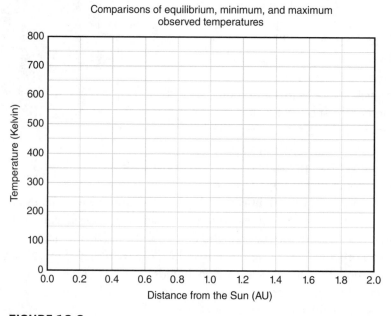

FIGURE 13.2

3. Compare the calculated equilibrium temperatures of Venus, Earth, and Mars. Remember that these temperatures include the effect of the distance of the planet from the Sun. Are these temperatures very different, or are they nearly the same, on this temperature scale?

4. Add the minimum observed temperature to the graph for each planet in Figure 13.2, using a small horizontal line.

5. Add the maximum observed temperature to the graph for each planet in Figure 13.2, using a small horizontal line.

6. Shade the area between the observed minimum and maximum temperatures that you have just graphed for each planet.

7. For which planets is there good agreement between the calculated equilibrium temperature and the range of observed temperatures? That is, for which planets does the range of observed temperatures include the calculated equilibrium temperature?

8. For which planets is there NOT good agreement between the calculated equilibrium temperature and the range of observed temperatures?

9. We made three assumptions when we calculated the equilibrium temperature. We assumed that
   a. the temperature of the planet is the same everywhere, which is clearly not true; planets will be hotter on the daytime side than the nighttime side.
   b. a planet's only source of energy is sunlight, which is nearly true for the terrestrial planets; the hot interior of Earth heats the surface less than 0.02 percent as much as the energy absorbed from the Sun.
   c. a planet is able to radiate freely into space; that is, it does not have an atmosphere.

10. Based on your analysis of the graph you just completed, which of the terrestrial planets would you guess has the most dense atmosphere? Which has the least dense atmosphere?

11. Consider your list of planets from question 7—those for which the calculated equilibrium temperatures agreed well with observations. What do the atmospheres of these planets have in common?

12. Consider your list of planets for which the temperatures did NOT agree. What do the atmospheres of these planets have in common?

13. Based on your answers to the last two questions, form a hypothesis about the effect of an atmosphere on the surface temperature of a planet. Finish this sentence: "If the temperature of a planet does not agree with the calculated equilibrium temperature, then . . ."

Step 3—Absorption versus Emission

14. Is Earth's atmosphere transparent or opaque in the visible part of the spectrum? State one observation you have made that demonstrates that your choice is true.

15. When visible light strikes a dark surface, it is absorbed, and the object heats up. State one example of an observation you have made that demonstrates that this is true.

16. The amount of visible light that is absorbed depends on how dark the object is. Objects that are close to black absorb nearly all the light that hits them, while objects that are nearly white reflect most of the light that hits them. State one observation you have made that demonstrates that this is true.

17. Much of Earth's surface is moderately dark. Extreme examples are roadways made of asphalt, which is particularly dark. The dark parts of Earth's surface absorb visible light. State one observation you have made that demonstrates that this is true.

18. Hot objects emit light. Relatively cool objects (a few hundred Kelvins, like a planet) emit primarily in the infrared. Relatively hot objects (a few thousand Kelvin, like a star) emit much more visible light. From this information combined with the preceding observations, we can draw the conclusion that the surface of Earth absorbs in the _____ part of the spectrum, but emits in the _____ part of the spectrum.

Step 4—Number versus Percentage

19. Astronomers put infrared telescopes in space. What can you conclude about Earth's atmosphere: Is it transparent or opaque in the infrared part of the spectrum? _____

20. Molecules absorb and emit light. This is, for example, why the sky is blue, as shown in **Figure 13.3**. Light comes into the atmosphere from the Sun, and molecules in Earth's atmosphere bounce the blue light around in the sky until it appears to come from some other direction. Which type of light, then, should take longer to reach the ground from space: yellow light or blue light?

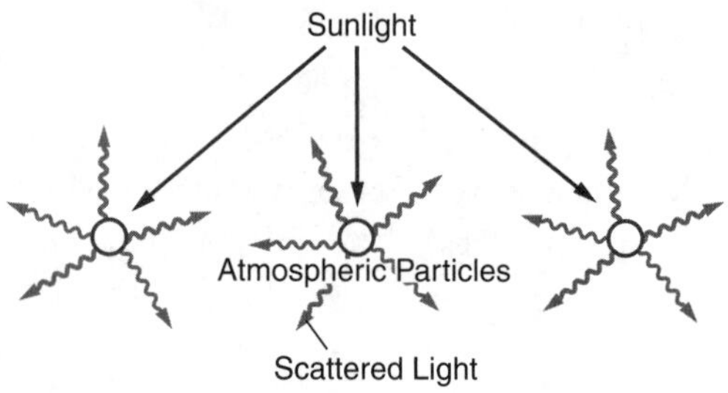

FIGURE 13.3

21. Molecules that absorb and emit infrared light are called greenhouse gases. As infrared light is emitted from Earth, what happens to it when it encounters a molecule of a greenhouse gas?

22. Consider two atmospheres that contain different amounts of greenhouse gas molecules. Which atmosphere will hold on to the infrared light longer: the one with more greenhouse gas molecules or the one with fewer? _____

23. Which planet will be warmer? The one that holds in the infrared light longer, or the one that lets it escape more quickly? This is the meaning of the term **greenhouse effect**.

Consider more carefully what is meant by "different amounts" of greenhouse gases. There are two ways to think about how much of something is in the atmosphere. You might consider the percentage: for example, the atmosphere is 95 percent carbon dioxide. Or you might consider the total number: for example, the atmosphere contains 100 million tons of carbon dioxide. How can we figure out which thing matters (the percentage or the number)? Consider Mars and Venus, which, as you have already seen, are two very different temperatures.

24. Consider Mars and Venus. Mars's atmosphere is 95.3 percent carbon dioxide. Venus's atmosphere is 96.5 percent carbon dioxide. How do these two percentages compare?

25. Mars's atmosphere is about 100 times thinner than Earth's, while Venus's atmosphere is about 100 times thicker. Which one has more carbon dioxide molecules? _____

26. What matters: the percentage of greenhouse gas molecules, or the number of greenhouse gas molecules? _____

Another way to think about that question is to consider an atmosphere made of one greenhouse gas molecule. That molecule can interact with only one infrared photon at a time. Doubling the number of molecules means that the atmosphere can now interact with two infrared photons at a time, doubling the effect. To think about the number of greenhouse gas molecules, we typically consider the **concentration**: We draw an imaginary box of a million atmospheric particles and ask how many of those molecules are carbon dioxide. (The vast majority of the other molecules are nitrogen, which is not a greenhouse gas.) Consider the graph in **Figure 13.4**, which shows how the number of carbon dioxide molecules in every "box" of atmosphere has grown in recent decades.

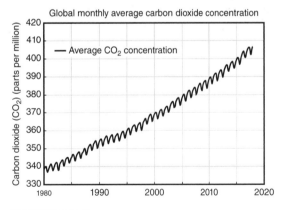

FIGURE 13.4

27. How much has the carbon dioxide concentration in Earth's atmosphere risen since 1980?

28. The preindustrial carbon dioxide concentration was 280 ppm. How much has the carbon dioxide concentration of Earth's atmosphere risen since preindustrial times?

29. Given how greenhouse gases affect temperature, what would you expect to see if you looked at graphs of temperature since 1980 and since preindustrial times?

30. Earth's global average temperature since 1980 is shown in **Figure 13.5**. It is shown as the "temperature anomaly," which tracks how different the current temperature is from a long-term average. Compare this graph to your prediction from the previous step. Were you correct in your prediction? Explain.

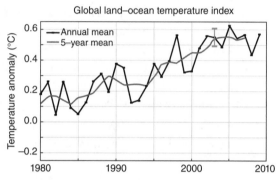

FIGURE 13.5

Step 5—Putting It Together

31. Using the key terms of this activity, describe why greenhouse gases in the atmosphere affect the temperature of a planet. Give examples from our own Solar System that show how we can explain the observed temperatures of planets through an understanding of greenhouse gases and greenhouse gas concentrations in the atmosphere.

● ACTIVITY 14
Comparing Storms on Jupiter and Earth

Learning Goals

Through the use of *Voyager I* spacecraft observations of the Great Red Spot (GRS) on Jupiter, you will not only learn about the wind conditions in Jupiter's upper atmosphere but also will

1. use reasoning to estimate the circumference of the GRS, an ellipse.

2. calculate the speed and rotation period of the GRS.

3. compare the results with one of the strongest western Atlantic hurricanes.

4. analyze why the lifetimes of the GRS and hurricanes are vastly different.

5. compare the speed of the GRS with published values.

6. discuss sources of uncertainties.

Key terms: local time, circumference, ellipse, uncertainty, rotation period, range of values

Step 1—Background

Before you begin, study the images taken by the *Voyager I* spacecraft from January 6 to January 29, 1979, shown in **Figure 14.1** in the appendix. These images were taken through a blue filter, one every Jupiter day as the spacecraft was approaching the giant planet. By taking pictures at the same Jupiter **local time**, the Great Red Spot (GRS) appears to remain stationary, while the belts and zones move across the image. Notice the action around the GRS as well as the belts and zones.

Follow the white feature (marked by an **x**) around one complete rotation on the outer edge of the GRS, as shown in Figure 14.1. There were 58 images taken over 23 Earth days. (We chose 11 of these for this activity.) Approximately how many hours, on average, passed between each image? We can calculate that:

$$\frac{23 \text{ days}}{58 \text{ images}} \times \frac{24 \text{ h}}{1 \text{ day}} = 9.5 \text{ hours between each image}$$

Two Earths would easily fit inside the GRS, as **Figure 14.2** shows. Calculating the circumference of an ellipse is hard, but we can make an estimate if we do some reasoning. The **circumference** of the Earth is about 40,000 km (*pi* × diameter). It appears the circumference of the **ellipse** outlining the GRS is a bit larger than 80,000 km, so we will add another 10 percent, making the circumference 88,000 km. Our **uncertainties** (our measurements are not exact) in this are about 10 percent, giving an estimate of 88,000 km ± 8,000 km. This is a common way for scientists to write a number with its uncertainty. The symbol "±" means "plus or minus"; "88,000 km ± 8,000 km" means that the most likely value is 88,000, but it might be as low as 80,000 (88,000 minus 8,000) or as high as 96,000 (88,000 plus 8,000).

FIGURE 14.2

1. From Step 1 information, what is the **rotation period** for the planet Jupiter? _____ hrs

Step 2—Calculating the Gas Speed of the GRS

To find the *speed* of the white feature, you will need a distance and a time. You have the distance it traveled around the GRS, but you also need to know how much time it took for that one rotation. We know that there were 9.5 hours in between images and there are 11 images total, so the total number of hours is just: 9.5 h/image × 11 images = 104.5 h.

2. Divide the circumference of the ellipse in kilometers by the number of total hours covered, to find the speed of the GRS at the distance of the white spot from its center. (Show your work here.)

_____ km/h

3. This is the approximate rotation *speed*. What is the rotation *period*, in days, at this distance from the center of the GRS? (**Hint**: What is the period in hours for the white feature?)

_____ days

Step 3—Comparison to One of the Strongest Atlantic Hurricanes

Since a couple of Earths "fit" into the GRS, it is meaningless to compare sizes for these two storms. However, we can compare a couple of other quantities: speeds and lifetimes.

On Monday, September 18, 2017, Maria made landfall on Dominica as a category 5 hurricane, as shown in **Figure 14.3**. Its maximum sustained winds were around 280 km/h (175 mph). It was the worst natural disaster on record to affect Dominica and Puerto Rico, as well as one of the deadliest Atlantic hurricanes, and in the top 10 most costly. Maria formed in the western Atlantic on September 16 and dissipated on October 2, 2017, for a time span of 16 days. Hurricanes can exist for as long as a month when traveling across the Pacific or Atlantic Oceans.

4. Quantitatively compare the maximum speed of Hurricane Maria to what you found for the GRS. Do so by taking a ratio to find out how many times faster the GRS was than Hurricane Maria. Show all logic and comment on the comparison.

$$\frac{\text{Speed of GRS}}{\text{Speed of Hurricane Maria}} =$$

FIGURE 14.3

5. Although the earliest observations of the GRS may date to the mid-1600s, records show that it has been continuously observed since 1830. Considering that the GRS is still present on Jupiter, how long has it been continuously observed?

_____ years

6. There are *many* things that are different between Jupiter and Earth, but what is so different that the GRS has lasted so much longer than any hurricane on Earth? Start by making a preliminary list comparing the planets, filling in the details for each planet.

PROPERTY	EARTH	JUPITER
Atmosphere		
Surface		
Interior		

7. In your opinion, which of these properties explains the difference in the extended lifetime of the GRS compared with any hurricane on Earth? Give the reasoning for your choice.

Step 4—Comparison with Published Values

8. Current measurements of the velocity of the outer parts of the GRS are 610 km/h [D.S. Choi et al., *Icarus*, 188:35–46 (2007)], with a given range of 430–680 km/h (see http://missionjuno.swri.edu/jupiter/great-red-spot). Quantitatively compare your results to the **range of values** from this reference and comment.

9. Review the uncertainties in the circumference, c, we used for the GRS and calculate the minimum ($c = 80{,}000$ km) and maximum ($c = 96{,}000$ km) speeds using the methods given in Step 2. The number of hours remains the same: 104.5 hours. With these new calculations, taking into consideration the uncertainties, does your comparison with the range of values from the professional reference change? How or why not? Show your work here.

Step 5—Putting It Together

10. We needed to make some estimates and assumptions for this activity that have affected our results, for better or worse. We also may not have considered everything related to the GRS. Regarding your comparison with the published measurements in question 8, identify what you think are *specific* sources of uncertainties. Be sure to use at least two of the key terms.

● ACTIVITY 15
Analyzing Saturn's Ring Particles

Learning Goals

In this activity, you will use Kepler's laws and Newton's law of gravity to analyze patterns in the rings of Saturn. After completing this activity, you will be able to

1. apply Kepler's third law to sections of the ring system of Saturn.

2. summarize the process by which a single moon will clear a gap in a ring system like Saturn's.

3. explain how the two shepherd moons manage to keep the particles that orbit between their orbits tightly confined within a narrow ring.

4. state how Newton's law of gravity affects the interactions among Saturn's moons and ring particles.

Key terms: accelerate, universal, shepherd moon, gap

Step 1—Background

Johannes Kepler developed his three laws based on observations of the orbits of planets in the Solar System. He found that describing the orbits as ellipses, not perfect circles, worked. He also noted that the closer a planet was to the Sun, the faster it moved. It didn't matter if this happened because the planet's orbit varied from a perfect circle—bringing it sometimes closer, sometimes farther from the Sun—or because the planet's orbit was at certain distance from the Sun.

We need Sir Isaac Newton's law of gravity to understand how objects interact with each other; Kepler's laws are incomplete in this sense. Newton's law says, "No matter what the mass of an object is, it exerts a force on other objects." This force depends on the product of the two masses divided by their distances apart, squared. The forces are equal and opposite; however, if the masses are different, the more massive object will **accelerate** the less massive one more.

Kepler's and Newton's laws are **universal**, which means they apply everywhere. We apply them to objects in the Solar System: planets, dwarf planets, comets, dust, and asteroids orbiting the Sun; moons and ring particles orbiting planets; and satellites orbiting Earth, Mars, Jupiter, and Saturn. They also apply to objects outside the Solar System.

Launched in late 1997, the *Cassini* orbiter satellite reached Saturn in early April 2004. It spent 13 years in the Saturnian system, sending back information that made us rethink much of our understanding of the Solar System. In mid-2017, scientists intentionally changed *Cassini*'s orbit, sending it closer and closer to Saturn until it finally plunged to its death in Saturn's atmosphere. After we take a quick look at Saturn, we will investigate the motions and interactions of the particles of its rings and what are called **shepherd moons** because they figuratively "guide" or "direct" ring particles in a particular direction.

There is a rule connecting what happens when an orbiting object loses energy (decelerates) or gains energy (accelerates): If an object in orbit is decelerated due to a force acting on it, it will move to a lower orbit. If an object in orbit is accelerated due to a force acting on it, it will move to a higher orbit. Objects then have the orbits corresponding to their new speeds.

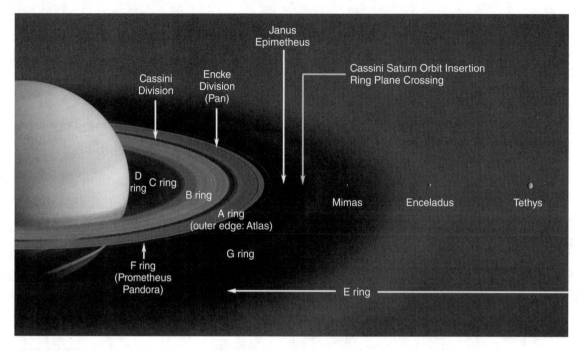

FIGURE 15.1

The image of Saturn and its icy rings and moons in **Figure 15.1** is an artist's concept of how the system fits together. The E ring, cut off at the right edge of the figure, extends almost twice as far as shown here. Titan and other moons are also farther away. The Encke Division (also called a **gap**) is caused by the moon Pan that orbits between the A and B rings. The F ring is kept narrow through the gravitational interaction of the moons Prometheus and Pandora. In this activity, you will investigate how the division and narrow ring are formed.

1. The *Cassini* orbiter satellite was torn apart by collisions with Saturn's rings.

Circle True or False

2. Saturn exerts a much greater force on the moon Pan than Pan exerts on Saturn.

Circle True or False

Step 2—Applying Kepler's Laws to Rings and Satellites

The rings of Saturn are made up of millions of icy chunks orbiting around it. An image taken by the *Cassini* spacecraft of a part of Saturn's rings is shown in **Figure 15.2**. These ring particles follow Kepler's laws.

From this image:

3. Which set of particles travels fastest? (Check one.)

A _____ B _____ C _____

4. Which set travels slowest? (Check one.)

A _____ B _____ C _____

FIGURE 15.2

The *Cassini* orbiter took hundreds of thousands of images of the moons and rings of Saturn. One containing the moons Janus, Pandora, and Prometheus, and the F and outer part of the A Ring of Saturn is shown in **Figure 15.3**. Saturn is below the image, out of the picture.

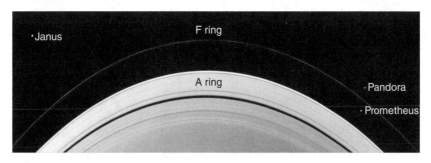

FIGURE 15.3

5. Rank the following objects in order of orbiting Saturn the fastest (1) to orbiting Saturn the slowest (4):

____Pandora   ____particles of the A Ring

____particles of the F Ring   ____Prometheus

6. Explain your logic in ranking the objects.

Let's apply Kepler's laws to something closer to home. In **Figure 15.4**, the space shuttle (a former NASA space ship) is depicted approaching a satellite in Earth's orbit. (Earth would be well below the bottom of the figure.) The shuttle has to capture the satellite to prevent it from falling out of orbit and burning up in Earth's atmosphere. The satellite is orbiting at a height of 250 km. The space shuttle is preparing to adjust its orbit in order to catch up to the satellite, and there are three possible approaches. There is approach A at an altitude of 200 km; B, at an altitude of 250 km; or C, at an altitude of 300 km.

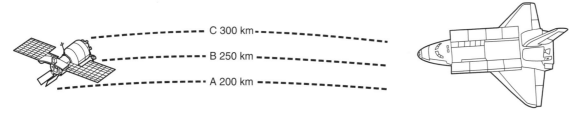

FIGURE 15.4

7. On which approach—A, B, or C—will the shuttle have the same speed as the satellite? _____

8. On which approach—A, B, or C—will the shuttle be traveling slower than the satellite? _____

9. That leaves a height of _____ where the shuttle will be traveling faster than the satellite.

10. A first-time commander of the space shuttle decides to orbit at 250 km and simply burn the shuttle's thrusters to catch up to the satellite rapidly. A mission specialist takes exception to this decision, stating that the space shuttle must catch up to the satellite from a lower orbit. Explain why the mission specialist is correct.

Step 3—How a Single Moon Clears a Gap in the Ring

Saturn has a number of gaps in its rings, some large, some small. How do the gaps form? A moon and two ring particles are shown in **Figure 15.5**. The inner ring particle will have a slightly higher speed than the moon, as indicated by the longer arrow in the direction of its orbit, and the outer ring particle will have a slightly slower speed than the moon, as indicated by the shorter arrow in the direction of its orbit. The moon will have a strong gravitational effect on both ring particles.

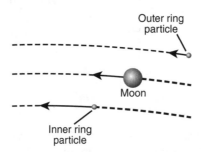

FIGURE 15.5

11. *Draw arrows* that represent the direction in which the moon's gravity will act on the two particles in Figure 15.5. Label these arrows to distinguish them from the arrows representing the speeds of the particles.

12. Just after a particle loses speed, will it move "down" or "up"; that is, closer or farther from the planet? _____

13. Conversely, just after a particle gains speed, will it move "down" or "up"; that is, closer or farther from the planet? _____

14. Will the inner ring particle accelerate (speed up) or decelerate (slow down) due to the interaction? _____

15. Will the outer ring particle accelerate or decelerate due to the interaction? _____

16. Now, *draw arrows* on Figure 15.5 that indicate the direction each particle will go in its orbit when affected by the single moon. Label these arrows to distinguish them from the arrows representing the gravitational forces.

17. As a particle falls "down" toward the planet, it will gain a little speed. This will stabilize it in a new, lower orbit. The outer particles, however, _____ speed as they move away from the planet, so they stabilize in a new, _____ orbit.

Step 4—How Two Moons Can Guide Ring Particles

In **Figure 15.6**, a ring is shown interacting with two moons. The inner shepherd moon will have a slightly higher speed and the outer shepherd moon will have a slightly lower speed than the ring particles, as indicated by the sizes of the arrows pointing in the direction of the moons' orbits. We once again need to consider Kepler's third law.

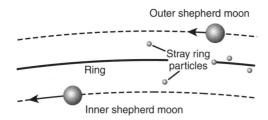

FIGURE 15.6

18. Do the stray ring particles move _____ **faster** or _____ **slower** than the outer shepherd moon?

19. Do the stray ring particles move _____ **faster** or _____ **slower** than the inner shepherd moon?

20. The moons have a strong gravitational effect on the stray ring particles. *Draw arrows* on Figure 15.6 that represent the direction in which each moon's gravity will act on the stray ring particle closest to it. Label these arrows to distinguish them from the arrows representing the speeds of the moons.

21. Now, consider your answers from the previous three questions, and *draw arrows* on Figure 15.6 that indicate the direction each stray ring particle will go in its orbit after being affected by the shepherd moon closest to it. Label these arrows to distinguish them from the arrows representing the gravitational forces.

Step 5—Putting It Together

22. Summarize the process by which a single moon will clear a gap in a ring system like Saturn's. Be sure to include Kepler's third law and Newton's law of gravity.

23. Summarize how the two shepherd moons manage to keep the particles that orbit between their orbits tightly confined within a narrow ring.

24. Suppose you are talking to a person with just a basic knowledge of how gravity works. Use Newton's law of gravity and how it affects Saturn's moons and ring particles to give that person more insight into orbits, forces, and acceleration. Be sure to use at least three key terms.

● ACTIVITY 16
Investigating Moons of the Giant Planets

Learning Goals

In this activity, you will compare seven major moons of the giant planets with Earth's Moon and with one another. As you work on this activity, you will

1. compare surface features of Jupiter's moon Ganymede with those of the Moon.

2. discuss evidence of resurfacing and cratering on four moons of Saturn.

3. relate characteristics of a lake on Saturn's moon Titan to a similar one on Earth.

4. examine the evidence for the reprocessing of part of Neptune's moon Triton.

5. compare features on Triton to ones on the Moon.

Key terms: comparative planetology, impact basin, resurfacing, saturation cratering, ejecta ray, tributary network

Step 1—Background

This activity continues our work in **comparative planetology**, a branch of science that compares characteristics among the bodies of the Solar System. Planetary scientists examine objects in the Solar System through ground-based observation, through satellites sent to planets, through analyses of meteorites—just about every way possible. The goal is to use the comparisons to determine how planets and other objects formed, the processes that shaped or are shaping them today, how they affect other objects, and how those comparisons relate to Earth. The main characteristics of the moons you will examine in this activity are gathered in **Table 16.1**. Take a moment to study these data, especially noting how the other moons compare to Earth's Moon.

From Table 16.1, you can see that the Moon's radius and mass are smaller than those of Ganymede and Titan, but its density is greater. That means that Ganymede and Titan must be more rocky and icy than the Moon. The table also gives a hint as to the lopsided satellite system of Saturn, where Titan dominates the mass of its other moons and rings. It also orbits much farther away. One of Neptune's moons, Triton, is about half the size of Titan, is less massive, but has roughly the same density. Information like this gives us a start in our analysis by asking: "What are the similarities? What are the differences? What do they mean?"

In **Figure 16.1** we have identified some of the large **impact basins** on Earth's Moon that were caused by large impactors billions of years ago. Sometime after the impact basins were created, lava flowed into them, essentially **resurfacing** that area. Those areas are younger than the area labeled **saturation cratering**, where if another impact were to occur, it would cover up previous impacts. The "ejecta rays" label is pointing to rays from the crater Copernicus. **Ejecta rays** are lighter lines of material that look like they are being shot out of a crater.

1. In Table 16.1, what trend can you find for radius and mass with distance from Saturn for its moons Mimas, Enceladus, Dione, and Rhea?

⊙ TABLE 16.1

Comparisons of giant planet moons to the Moon

PLANET	MOON	SEMIMAJOR AXIS (10³ KM)	RADIUS (KM)	MASS (10²⁰ KG)	DENSITY* (KG/M³)
Earth	Moon	384	1,738	735	3,340
Jupiter	Ganymede	1,070	2,631	1,482	1,940
Saturn	Mimas	186	198	0.4	1,150
	Enceladus	238	252	1	1,600
	Dione	377	562	11	1,480
	Rhea	527	764	23	1,230
Saturn	Titan	1,222	2,575	1,346	1,880
Neptune	Triton	355	1,353	214	2,060

*Water = 1,000 kg/m³

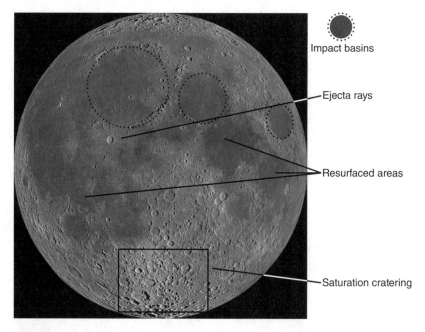

Impact basins

Ejecta rays

Resurfaced areas

Saturation cratering

FIGURE 16.1

Step 2—Jupiter's Moon Ganymede

Start with a comparison between the Moon and the largest moon in the Solar System, Jupiter's moon Ganymede, as seen in **Figure 16.2**. The images have been resized to show the scale between their actual sizes. Look for features given in Figure 16.1.

2. Describe two surface features that Ganymede appears to have in common with the Moon.

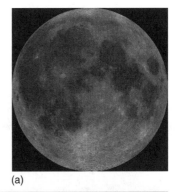

(a)

3. How are these features similar? Consider the properties of the Moon and Ganymede. In what ways must the origins of the features on Ganymede be different?

(b)

FIGURE 16.2

Step 3—Four Satellites of Saturn

As of 2018, Saturn has at least 62 known moons. We show four of these in **Figure 16.3**: Rhea, Dione, Enceladus, and Mimas. Because the images have been resized to all be the same size, we have listed their radii. Note that Rhea is almost four times larger than Mimas. Study these images and note the amount of cratering, unusual terrain, shading, and other features of each moon.

(a) Rhea, radius = 764 km   (b) Dione, radius = 562 km   (c) Enceladus, radius = 252 km (d) Mimas, radius = 198 km

FIGURE 16.3

4. Which moon(s) experienced resurfacing? Describe the evidence.

5. Are there areas of saturation cratering on any of these four moons? Which one(s) and on what part?

6. Rank these moons in order of the one having the highest number of craters on the side shown to the one having the lowest number.

    Highest _____ _____ _____ _____ Lowest

7. Mimas and Enceladus orbit within Saturn's outer rings, while Dione and Rhea orbit beyond the major rings. Could that explain the differences in cratering? Why or why not? (Look for exceptions in patterns.)

8. The particles of the E ring are extremely small, more like fine icy mist than crushed ice. Theory states that those particles would rapidly dissipate unless they were being replenished. Observations show that Enceladus orbits within the E ring. Formulate a testable hypothesis as to why the particles of the E ring exist by completing this sentence: "If Enceladus is responsible for replenishing the particles of the E ring, then . . ."

Step 4—Saturn's Moon Titan and Neptune's Moon Triton

Titan

Even though Titan is a moon and is about 1.5 times the radius of the Moon, it is more interesting to compare Titan's surface features with those of Earth. Earth's average surface temperature is about 290 K; Titan's is about 94 K. The *Cassini* spacecraft, which orbited Saturn from 2004 to 2017 and passed by its moons many times, used infrared instruments and radar to map most of Titan's surface. There are a large number of features that have analogies on Earth, including a nitrogen atmosphere. We investigate the most intriguing feature, a lake with **tributary networks** that resemble tributary networks on Earth as brooks flow into streams that flow into rivers.

9. The images in **Figure 16.4** are from two different worlds. Figure 16.4a shows Lake Mead, the reservoir behind the Hoover Dam on the Colorado River, while Figure 16.4b shows Ligeia Mare (mare is Latin for "sea") located on Titan. Examine the pair of images shown in Figure 16.4. Compare/contrast characteristics along their shorelines and surrounding area. List at least two things that look very similar and at least one thing that is quite different.

10. Consider the characteristics of the surfaces of these two worlds. Could the features on Titan be due to liquid water flowing into a lake? Explain.

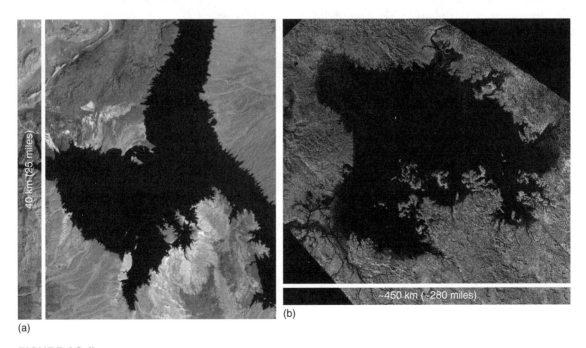

(a)

(b)

FIGURE 16.4

Triton

Neptune and its moon Triton have had only one visitor from Earth: The *Voyager 2* flyby spacecraft in 1989. Triton's surface held a number of surprises. *Voyager 2* found evidence of reprocessing of Triton's surface, shown in the top half of **Figure 16.5**, along with what appeared to be nitrogen ice geysers (the dark streaks in the lower part).

11. Look carefully at this *Voyager* image and describe the two different regions of Triton. What distinct features do you see in each region?

FIGURE 16.5

The image of Triton in **Figure 16.6** shows a plain of ice. The plain was probably formed by eruptions of water or water-ammonia slurry. It seems to fill the remains of ancient impact basins.

12. What region(s) on Earth's Moon look like they might have similar origins?

13. Why might there be differences between the formation of this plain on Triton and those similar features on Earth's Moon? Consider aspects of where they are located, their sizes, and their densities.

FIGURE 16.6

Step 5—Putting It Together

14. Explain what is meant by comparative planetology by summarizing the comparisons you made and the conclusions you reached in this activity. The key terms should be used as a guideline for what to summarize.

● ACTIVITY 17
Classifying Meteorites

Learning Goals

In this activity, you will learn about different types of meteorites and their origins. As you work on the activity, you will

1. examine images of samples of meteorites and note dominant features and possible classification.

2. distinguish among iron, stony-iron, and stony meteorites.

3. classify "unknown" meteorites using supportive explanations.

4. compare spectra of an asteroid and a meteorite and evaluate the evidence for the origin of the meteorite.

Key terms: meteorite, meteoroid, meteor, fusion crust, differentiate, iron, stony-iron, stony, reflectance spectra, chondrule

Step 1—Background

Meteorites are rocks from space. They have traveled through space as **meteoroids**, created the blazing trail of a **meteor** as they fell through the atmosphere, and survived to reach the ground as a solid object known as a meteorite. As it passes through the atmosphere, the meteoroid develops a smooth outer layer called a **fusion crust**. If an asteroid is quite large, it will heat up significantly during formation as it contracts or is bombarded by other asteroids. If it becomes molten, it may **differentiate**: heavy elements like **iron** may fall to the center, while light elements found in stone, like silicon, remain closer to the outside. If a meteorite came from the central part of a differentiated asteroid, it may be almost entirely iron. If a meteorite came from the outer part of a differentiated asteroid, it may be a **stony-iron** or entirely **stony**. Stony meteorites are classified as either chondrites or achondrites. Carbonaceous chondrites have high percentages of volatiles and may be from objects that have never been heated.

How do we know that meteorites come from space? One reliable piece of evidence suggests that at least some originated from collisions in the asteroid belt. Many observations from a number of widely spaced locations on Earth of a meteor falling through Earth's atmosphere are used to determine the meteorite's origin. Another piece of evidence comes from comparing **reflectance spectra** of asteroids and meteorites. Reflectance spectroscopy uses the methods of spectroscopy to study objects that reflect the Sun's light rather than produce their own light. If the chemical composition of an asteroid and a meteorite are the same, then the reflectance spectra should be very similar.

Meteorites often contain **chondrules**—round grains that form as droplets in space before being accreted into an asteroid. These are some of the oldest solid materials in the Solar System. The surface of a meteorite is often a fusion crust made of melted and warped rock that was heated by friction as the meteoroid fell through Earth's atmosphere.

1. Suppose that a meteorite comes from the boundary between the differentiated iron core and the stony outer crust. What would you expect to observe about its composition?

Step 2—Studying Meteorite Images

Study **Figure 17.1** in the appendix. Six examples of four kinds of meteorites are shown. Examine each image carefully. Pay particular attention to the fusion crust (if any), chondrules (if any), and metallic characteristics (if any). Fill in **Table 17.1** with as much detail as you can. Be as specific as possible. Use your newly found knowledge to classify some "unidentified" meteorites shown in **Figure 17.2** in the appendix. Place your classifications in **Table 17.2**.

● TABLE 17.1

Notable characteristics of the six examples of meteorites shown in Figure 17.1

	IDENTIFICATION AND LOCATION OF FIND	DOMINANT FEATURES	CHONDRULES? (DESCRIBE)	METALLIC? (DESCRIBE)
(a)	**Carbonaceous Chondrite** Pueblito de Allende, Chihuahua, Mexico			
(b)	**Chondrite** Cocklebiddy, Western Australia			
(c)	**Iron** Wolfe Creek, Mataranka, Northern Territory, Australia			
(d)	**Iron** Great Namaqualand, Namibia, Africa			
(e)	**Stony-iron** Brenham meteorite, Haviland, Kansas			
(f)	**Achondrite** (Eucrite) Millbillillie, Western Australia			

● TABLE 17.2

Classification of "unidentified" meteorites shown in Figure 17.2

	TYPE	REASON FOR CLASSIFICATION
(a)		
(b)		
(c)		
(d)		

Step 3—Drawing Conclusions

2. The classification of "stony" that we use includes chondrites (carbonaceous chondrites, a special kind) and achondrites (without chondrules). The stony meteorites in this activity probably came from the surface of a differentiated asteroid. What does "differentiated" mean?

3. Why are smaller, rocky bodies in the Solar System (say, 1,000 meters in diameter or less) generally not differentiated, while larger bodies are?

4. Explain why we believe iron meteorites must come from large asteroids.

The graph in **Figure 17.3a** compares the reflectance spectrum of an asteroid from the asteroid belt (dots) and the reflectance spectrum of some of the small grains of the Millbillillie achondrite meteorite (solid line), shown in **Figure 17.3b**.

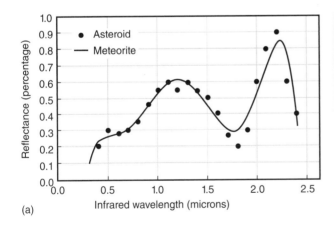

(a)

(b)

FIGURE 17.3

5. Is the spectral evidence solid enough to state that this meteorite was once part of the asteroid that was observed? Support your answer.

6. Suppose that you discover a meteorite and take its reflectance spectrum. You find that the reflectance spectrum has a broad peak of about 90% reflectivity from 0.4 to 0.6 microns, dips sharply to 70% at 0.9 microns, and rises again to flatten out and be approximately constant at about 80% reflectivity from 1 to 2.5 microns. Sketch the spectrum of the meteorite you have discovered onto Figure 17.3a. Could this meteorite have come from the same asteroid as the Millbillillie meteorite shown in Figure 17.3b? Support your answer.

Step 4—Putting It Together

7. Assume that the iron, stony-iron, and stony meteorites you examined in this activity all came from a particular asteroid at different times, and that this asteroid has been identified by scientists. Using at least three key terms, describe a possible scenario for the history of these meteorites over the past 4 billion or so years, up through their discovery and identification.

● ACTIVITY 18
Spectral Classification of Stars

Learning Goals

In this activity, you will learn to determine fundamental properties of stars through a study of their spectra. By working through this activity, you will be able to

1. apply Wien's law to real spectra and estimate the surface temperatures of three stars.

2. classify stars by their surface temperatures.

3. describe in detail the characteristics of sample stellar spectra and the dominant elements showing absorption lines.

4. state the reason why very hot and very cool stars show weak hydrogen lines in their spectra.

Key terms: electromagnetic spectrum, blackbody, peak wavelength, infrared, surface temperature, spectral type, Wien's law, stellar atmosphere, absorption line, photon, ionized, Balmer lines

Step 1—Background

As radiation makes its way out of a star, it becomes spread out across all the colors of the visible rainbow and beyond. The star will be brighter in some parts of the **electromagnetic spectrum** than others. The graph of the spectrum of a star has a particular shape that is associated with objects that shine because they are hot, rather than for some other reason such as reflection or electrons bouncing up and down in atoms. A spectrum of this particular shape is called a **blackbody** spectrum. Examples of this shape are shown as dotted curves in **Figures 18.1, 18.2,** and **18.3.**

The blackbody spectrum has one wavelength at which it is brightest, called the **peak wavelength.** This peak in the spectrum might be outside of the part of the spectrum that you can see with your instrument. For example, in Figure 18.3, the peak lies to the right of the visible part of the spectrum shown. In this case, the peak is in the **infrared** part of the spectrum. The peak wavelength depends on the **surface temperature** of the star—that is, the temperature in the layer that marks the boundary where the star becomes transparent. Stars are grouped into **spectral types** according to their temperature, which can be calculated from the peak wavelength. **Wien's law** states that the peak wavelength of a blackbody spectrum, λ_{peak}, in nanometers (nm) (1 nm = 10^{-9} meter), is inversely proportional to its surface temperature, T, in kelvin:

$$\lambda_{\text{peak}} = \frac{2.9 \times 10^6}{T}$$

1. Solve for T, and show the equation here: _____

The outer layers of a star make up its **stellar atmosphere.** The atoms in these layers affect the radiation from the star as it flows outward. Atoms in these atmospheres absorb light at particular wavelengths, creating **absorption lines** in the star. These are observed as dips in the spectrum (often quite narrow), where the spectrum drops below the blackbody curve. Each type of atom in the periodic table of the elements produces absorption lines at particular wavelengths, so identifying the absorption lines that are present in the spectrum tells you what atoms are present in the atmosphere of the star.

2. Look at the spectrum in Figure 18.1, which shows the wavelengths of absorption lines belonging to particular atoms with vertical dotted lines. The atom names are abbreviated as one or two letters plus a roman numeral at the top of the dotted line. List three different types of atoms that are present in this star.

Absorption lines are caused by electrons in an atom that absorb radiation. The absorbed energy causes the electron to jump up from one energy level to another. Electrons in an atom can only occupy particular energy levels, so the amount of energy needed to jump from one level to another has only specific values. The energy in a packet of radiation (called a **photon**) is related to the wavelength of the photon. Therefore, electrons cannot absorb just any old photon; they can only absorb photons with the right wavelength to give the electron the right amount of energy to step up to another energy level. Each type of atom has unique absorption lines because each type of atom has a unique set of energy levels that electrons can occupy. The only exception is if the absorbed photon has so much energy that the electron leaves the atom entirely. The atom, lacking an electron, has been **ionized**. The electron carries away any extra energy as kinetic energy.

Imagine a person on a set of stairs. The person can jump up one step, or two steps, but she must always jump from the step she is currently on. Just as a person can only jump from the step she occupies, an electron can jump only from the energy level it occupies. If an electron is in energy level 1, it cannot jump out of energy level 2. If an electron is in energy level 3, it cannot jump out of energy level 2. For example, in very hot stars, nearly all of the hydrogen atoms are ionized, so very few of their electrons can jump out of energy level 2. In very cold stars, nearly all of the hydrogen atoms have their electrons in energy level 1, so very few of their electrons can jump out of energy level 2. Stars that are in between very hot and very cold have lots of atoms with their electrons in energy level 2, so lots of electrons can jump out of this level. Electrons in hydrogen atoms that jump out of energy level 2 absorb light in the visible part of the spectrum, so this is a useful set of absorption lines that are easy to observe in the visible part of the spectrum. They are so useful, in fact, that this group of absorption lines caused by an electron jumping out of energy level 2 is known as the **Balmer lines** of hydrogen.

3. Explain why the Balmer lines would be weak in a very hot or very cold star, but strong in a star with an intermediate temperature.

Step 2—Finding the Spectral Type of Stars

4. For the stars in **Figures 18.4, 18.5,** and **18.6** (stars II, III, and V), visualize and then draw the blackbody spectrum that entirely fits over it. A dotted curve above the spectrum in Figures 18.1, 18.2, and 18.3 illustrates how this is done. Find the peak wavelength of each stellar spectrum by reading the wavelength on the *x*-axis directly below the peak in the spectrum and enter it in **Table 18.1**.

● TABLE 18.1

Characteristics of six SDSS stars based on examination of their spectra

STAR ID	PEAK WAVELENGTH	SURFACE TEMPERATURE	SPECTRAL TYPE
I	<300 nm	>10,000 K	B
II			
III			
IV	about 480 nm	~6000 K	F or G
V			
VI	>920 nm	<3100 K	M

● TABLE 18.2

The spectral types associated with each surface temperature range for stars

SPECTRAL TYPE	TEMPERATURE RANGE (K)	SPECTRAL TYPE	TEMPERATURE RANGE (K)
O	>33,000	G	5200–6000
B	10,000–33,000	K	3700–5200
A	7500–10,000	M	<3700
F	6000–7500		

5. Using the relationship between temperature and peak wavelength that you found in question 1, find the temperatures of the three unclassified stars. Enter your values in Table 18.1. Notice that Table 18.1 lists the stars in order by temperature, starting with the hottest one. This will help you check your math.

6. Use the information in **Table 18.2** to find the spectral type of stars II, III, and V. Enter the type of each star into Table 18.1.

Step 3—Further Investigations of the Spectra

7. Study the comments accompanying Figures 18.1, 18.2, and 18.3, and then study the spectra of the last three stars. Add comments for the last three stars in the space provided.

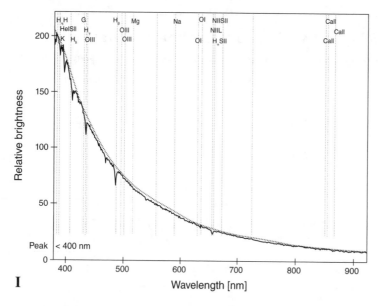

FIGURE 18.1

Star I is the hottest star in our sample. Its peak wavelength lies at a much shorter wavelength than 400 nm. We included additional criteria for this statement: the hydrogen lines are extremely weak.

We calculated its surface temperature at 100 nm and then assumed it is much hotter than that. Doing so, we find:

$$T \geq 29,000 \text{ K}$$

Its spectral type is B, according to Table 18.2.

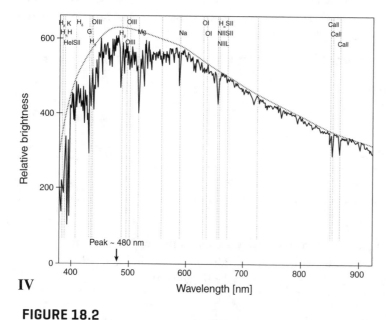

FIGURE 18.2

Star IV has a peak wavelength estimated at about 480 nm. We have drawn a smooth curve over the shape of the spectrum, showing what a blackbody or thermal radiation curve looks like in general.

This peak wavelength gives us a surface temperature of ~6000 K.

The spectral type is F or G.

Learning Astronomy by Doing Astronomy Second Edition

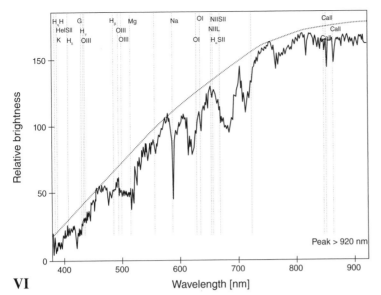

VI

FIGURE 18.3

Star VI is the coolest star in our sample. It shows lots of absorption lines and some deep absorption features that span a range of wavelengths. It is the deep absorption features that indicate molecules are present in its atmosphere. We probably do not see its peak wavelength and so assume it is at a wavelength longer than 920 nm.

If the peak were 920 nm, then the surface temperature would be 3100 K.

Its spectral type is M.

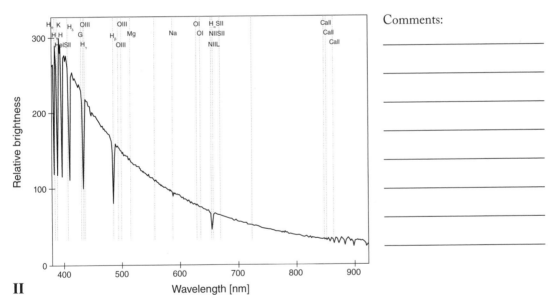

II

FIGURE 18.4

Comments:

_____

_____

_____

_____

_____

_____

_____

_____

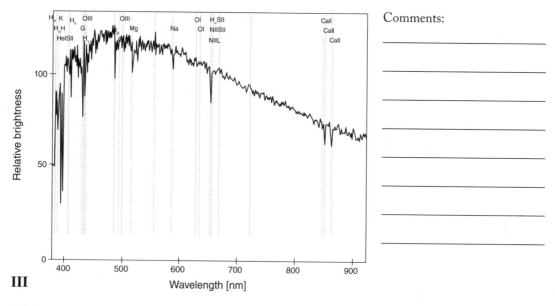

III

FIGURE 18.5

Comments:

_____

_____

_____

_____

_____

_____

_____

_____

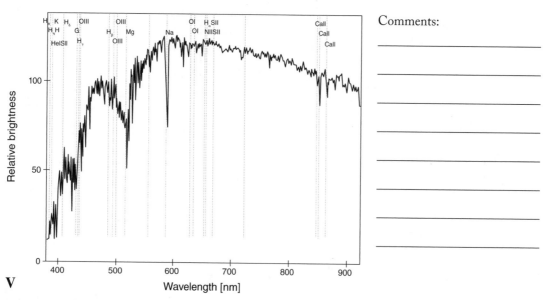

V

FIGURE 18.6

Comments:

_____

_____

_____

_____

_____

_____

_____

_____

Step 4—Putting It Together

8. Using key terms from the activity, generally describe how the spectra of these stars change in appearance from the hottest star (I) to the coolest star (VI). List the types of atoms that show up prominently in hot versus cold stars. Discuss the overall shapes of the spectra, the location of the peaks, and how and why the absorption lines differ among the stars in this sample.

● ACTIVITY 19
Finding Distances to Stars Using Parallax Measurements

Learning Goals

In this activity, you will determine a relationship between distance and apparent motion of a nearby object when viewed from two vantage points. You will apply this relationship to the measuring of distances to stars. As you work through this activity, you will

1. qualitatively state relative distances to stars based on their parallax shift.

2. describe the limits of the measured parallax method of determining distances to stars.

3. apply this knowledge to the measured parallax angles of stars.

Key terms: parallax, parallax angle, baseline, astronomical unit (AU), inversely proportional, arcsecond (arcsec), parsec (pc)

Step 1—Background

Parallax is a method of measuring the distance to an object by measuring a **parallax angle** from two different observation points (this is also sometimes called triangulation). Astronomers use this method to find the distance to nearby stars. They make observations at two different points in Earth's orbit and observe how nearby stars appear to move against the background pattern of stars that are much farther away. The distance between the observation points is called the **baseline**. A longer baseline is better: If you make the baseline between observations twice as long, you can measure distances that are twice as far away. For astronomers, this generally means that they try to observe the same star from opposite sides of Earth's orbit (in January and July, for example).

You can see the parallax effect using your thumb. Hold your thumb out at arm's length, look at a distant object, and alternately open and close each eye. Your thumb will seem to jump back and forth relative to the background. This is because the baseline between the centers of your eyes is about 7 centimeters (cm), so each eye has a slightly different point of view. Bring your thumb closer to your eyes and try the experiment again. Did your thumb appear to move more or less when it was closer?

1. ___X___ **True** or ____ **False**: If we had a satellite orbiting at the distance of Mars (about 1.5 AU from the Sun), we could measure distances with similar accuracy 1.5 times farther away.

Step 2—Parallaxes of Stars

Astronomical parallax is measured as shown in **Figure 19.1**. The length represented by the letter B indicates the diameter of Earth's orbit. At right is a star at great distance, d. The angle, α, is a measurable quantity, which is equal to twice the parallax angle, p. The baseline for the measurement of parallax angles for stars is the diameter of Earth's orbit. Because the parallax angle is half of α (the thing we can measure), we use the radius of Earth's orbit for the baseline. The radius of Earth's orbit is 1 **astronomical unit (AU)**.

All stars except the Sun are so far away that this angle is very small. For small angles like this, we could use trigonometry to find that the parallax angle, p, is **inversely proportional** to the distance, d; if the distance is small, the parallax angle is large, and vice versa. In other words, if star A has 1/4 the parallax angle of star B, star A is four times as far away as star B. Once we know the parallax angle and the baseline, we can find the distance to the star.

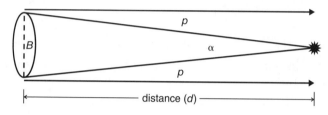

FIGURE 19.1

2. Imagine that you measure the parallax of two stars in the constellation Leo. Regulus has one-half the parallax angle of Denebola. What do you immediately know about the relative distances of these stars from Earth?

3. What if there were a star that was 1,500 times farther away than the star Regulus? Assuming no advancement in our technology, would we still be able to detect an apparent shift in its position between January and July?

4. The limit to our measurements of parallax angles is about 0.005 arcseconds, or about 650 light years. The diameter of the Milky Way is about 100,000 light years. To what fractional part of the diameter of the Milky Way can we determine distances using this method?
   a. 0.0065   **b.** 0.65   **c.** 154   **d.** 0.154

5. Suppose astronomers put a telescope on Pluto, ~40 AU from the Sun, beaming back information to us. How much *farther* will they be able to measure accurate parallaxes compared with their work here on Earth? (**Hint**: Did you read all of Step 1?)

6. What do your answers to questions 3–5 tell us about the limits of using measured parallaxes of stars to find their distances?

Step 3—Introducing the Parsec

It is convenient to define a new unit that is connected to the way we measure parallax. By definition, a star that has a parallax angle of 1 **arcsecond** has a distance of 1 **parsec**. A star that is 2 parsec (pc) away has a parallax angle of 1/2 arcsecond (arcsec; see **Figure 19.2**). A star that is 5 parsecs away has a parallax angle of 1/5 arcsecond, and so on. The unit "parsec" makes converting from parallax angle to linear distance straightforward; simply divide 1 by the parallax angle to get the distance ($d = 1/p$). Once again you see that the distance and the angle are inversely proportional; if the angle is large, the distance is small.

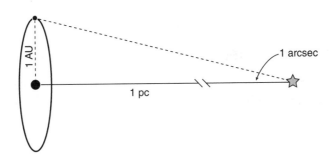

FIGURE 19.2

7. Rank the stars in **Table 19.1** in order of distance from Earth.

● TABLE 19.1 Rank: closest = 1, farthest = 4		
RANK	**STAR NAME**	**PARALLAX ANGLE (ARCSEC)**
	Antares	0.024
	Ross 780	0.213
	Regulus	0.045
	Betelgeuse	0.006

8. Calculate the distances to the stars listed in **Table 19.2**. The parallax angle is given in arcseconds, and the distance will come out in parsecs.

◉ TABLE 19.2

Calculating distances to stars

STAR	PARALLAX ANGLE (ARCSEC)	DISTANCE (PARSECS)
Arcturus	0.090	
Procyon	0.288	
Hadar	0.006	
Altair	0.194	

Step 4—Putting It Together

9. Briefly explain, using key terms from this activity, how astronomers calculate the distance to a star using parallax. Are there stars for which this method cannot be used? Explain.

● ACTIVITY 20
Analyzing a Solar Coronal Mass Ejection

Learning Goals

In this activity, you will take a close look at a coronal mass ejection (CME) from the Sun. After completing the activity, you will be able to

1. calculate the speed of the material ejected.

2. determine how much advanced warning we would have.

3. consider the source(s) of errors and uncertainties.

4. summarize why it is important for us to predict CMEs accurately.

Key terms: coronal mass ejection (CME), plasma, image scale, uncertainty

Step 1—Background

Coronal mass ejections (CME) are blobs of very hot material that escape from the Sun's corona, usually when a loop of magnetic field snaps. These blobs are so hot that the material is completely ionized, forming a **plasma** in which the electrons and the atomic nuclei are separated and can be treated as independent gases. If the CME is released in the right direction to strike Earth, these fast-moving charged particles can interact with Earth's magnetic field, causing beautiful aurorae. They can also interfere with the operation of sensitive electronics in orbit around Earth, much like static electricity from shuffling across a carpet can "fry" your electronics down here on the ground.

1. List one positive effect of CMEs on Earth, and one negative effect.

Step 2—Finding the Speed of a CME

2. Study **Figure 20.1** in the appendix and identify the "clump" of material that can be traced over a series of images.

3. Find the center of the clump in each image, and measure how far it has traveled away from the Sun. Record these data into **Table 20.1**. The ruler that has been overlaid on each image will help. Notice that the markings on the ruler are in millimeters.

⬤ TABLE 20.1

Measurements of the distance traveled by the coronal mass ejection from the Sun

TIME (UT)	HOURS	SCALED DISTANCE ON IMAGE (MILLIMETERS)
10:45	0	7
11:42	1	10
12:42	2	15
13:42	3	20
14:42	4	25
15:42	5	30

4. Graph your data in **Figure 20.2**. Draw a straight line that follows the trend in the data. This is your approximate best-fit line.

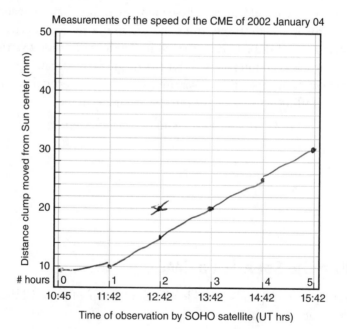

FIGURE 20.2

5. Find the slope of the line by calculating the "rise over the run." (Here's a reminder if you've forgotten how to find the slope: divide the difference in *y*-positions of two points on the line by the difference in *x*-positions. Don't forget to use the values from the axes, rather than just counting tick marks on the graph.) Your answer will come out in millimeters per hour (mm/h). What is the slope of the line in millimeters per hour?

Slope = _____ mm/h
         0.75

6. Calculate the image scale: The small central circle shown in white represents the size of the Sun in the image. The actual diameter of the Sun is approximately 1.4×10^6 kilometers (km). If we measure the diameter of the small central circle, we can find out how many km are represented by a mm in the image. This is the **image scale**. For example, if the small central circle measured 7 mm across, then in this image, every 7 mm represents 1.4×10^6 km per 7 mm, or

$$\frac{1.4 \times 10^6 \text{ km}}{7 \text{ mm}} = \frac{2 \times 10^5 \text{ km}}{1 \text{ mm}} = 2 \times 10^5 \text{ km/mm}$$

a. Measure the diameter of the circle that represents the Sun in millimeters.

Diameter of the circle = _____ 12 _____ mm

b. Divide the actual diameter of the Sun by your value for the diameter of the circle in mm to find the image scale.

Image scale = _____ 116,666.667 _____ km/mm

7. Multiply your slope (question 5) by the image scale to find the average speed (in km/h) of the CME over the duration of the observations.

Average speed = _____ 87500 _____ km/h

Step 3—Finding the Time It Takes for the CME to Arrive

8. This particular CME was not traveling toward Earth. Others, however, occasionally do. The distance between Earth and the Sun is about 1.5×10^8 km. Dividing this distance by the speed of the CME will give the time it takes the CME to arrive at Earth. This is how much advance notice scientists could have that a CME is coming. Consider a CME directed at Earth, traveling at the average speed you calculated in Step 2.

How long would it take this CME to reach Earth? _____ 1714.28 _____ hours

Convert that time from hours to days. _____ 71.4 _____ days

Step 4—Considering Uncertainties

9. Your results depend primarily on accurately measuring the Sun's diameter, locating the same clump in each of the images, and accurately determining the center of the clump as the material spreads out. Let's consider just the accuracy of the measurement of the Sun's diameter on the image and find out how an **uncertainty** in that measurement propagates along through the calculation of the average speed. Calculating uncertainties is one of the most important steps in any scientific project; it's important to know how far wrong you could be and still be right!

One way to calculate this is to estimate your uncertainty, and then repeat all your calculations using the highest possible value and also the lowest possible value. In the end you will have a "most likely" value, as well as a range of values that are all reasonable, given the limitations on the accuracy of your measurements. Here, we work through the calculation of uncertainties with you, using our measurement of 7 mm for the diameter of the circle in the image. Follow along using your measurement from 6a.

a. What is your original measurement for the diameter of the Sun (from 6a)?

_____ 12 _____ mm

b. Suppose our measurement of the circle diameter (7 mm) was accurate only to 25 percent. Twenty-five percent of 7 mm is just about 1.8 mm. The largest diameter that is still reasonable with this level of accuracy is 7 mm + 1.8 mm = 8.8 mm; the smallest is 7 mm − 1.8 mm = 5.2 mm. (Scientists often compress this information by writing 7 mm ± 1.8 mm, to show that we could add or subtract the 1.8 mm and that would define our reasonable range.) Assume that your measurements from 6a are also accurate only to 25 percent. What would be your smallest and largest reasonable diameters for the circle?

   i. Smallest: _____9_____ mm   ii. Largest: ____15____ mm

c. In Step 2, we divided the actual diameter of the Sun by the diameter of the circle in the image to find the image scale. Carrying through this operation for both the smallest (5.2 mm) and largest (8.8 mm) reasonable diameters gives a largest reasonable image scale of 269,000 km/mm and a smallest reasonable image scale of 160,000 km/mm. Repeat this division two times to find your smallest and largest image scale based on your measurement of the circle diameter.

   i. Smallest: ~~15555~~ 93 333 km/mm   ii. Largest: __155555__ km/mm

d. Take the last step on your own. To find the speed in km/h, you multiplied the slope from question 5 by the image scale. Repeat this calculation using your smallest and largest image scale to find the smallest reasonable speed and the largest reasonable speed of the CME.

   i. Smallest: __69999__ km/h   ii. Largest: __116 666__ km/h

10. Coronal mass ejections have a large range of velocities, with an average speed being around 500 km/s (1,800,000 km/h). How does the value you found for the speed of this CME compare to the average? Does this average value lie within the range of your uncertainties? Be quantitative by finding the percent difference between your measured value and the published average speed. (Show all logic, and comment on your results.)

$$\frac{(your\ value) - 1.8 \times 10^6}{1.8 \times 10^6} \times 100\% = \text{percentage error}$$

Step 5—Putting It Together

11. Using key terms of the activity, explain why it is important for scientists to calculate uncertainties, both as a general principle and for this particular situation of CMEs headed toward Earth. Why would it be important to know not only the most likely speed of the CME, but also the slowest and fastest speeds it might reasonably be traveling?

● ACTIVITY 21
Understanding the Evolution of the Sun

Learning Goals

After working through this activity, you will be able to

1. state where fusion is occurring at a given evolutionary stage.

2. analyze what is supporting the core and the rest of the Sun at each stage.

3. determine if the inward force of gravity and the outward pressure are balanced.

4. determine what will happen to the core and the rest of the Sun if the Sun is not in equilibrium.

5. summarize the eventual fate of the Sun.

Key terms: pressure, ideal gas law, kinetic energy, radiative energy, electron degeneracy, greater than (>)

Step 1—Background

There are some important rules that govern the evolution of the Sun, from now until its end:

- Gravity will compress matter into the tiniest sphere possible unless some opposing force prevents it. **Pressure** is the force per unit area applied in a direction perpendicular to the surface of an object. (Think of the pressure in the tires of a bike or a car: "PSI" means "pounds per square inch," which is a measure of pressure. It pushes outward on the walls and treads of the tire.)

- Regions of the Sun obey the **ideal gas law**, which states that the pressure (P), volume (V), and temperature (T) of a gas are related to each other: $P \propto T/V$. If the temperature goes up or down while the volume stays the same, then the pressure will increase or decrease. If the pressure is held constant, an expanding gas will cool down and a contracting gas will heat up.

- Outwardly directed pressure within a region of a star is provided by the motion of particles in it (thermal pressure due to the **kinetic energy** of the particles) and by the photons passing through it (**radiative energy**).

- Electrons exert a different kind of pressure, **electron degeneracy**, when they can't be "pushed" any closer together, according to the rules of quantum mechanics.

- Whether or not fusion occurs in the core of a star depends on the temperature there.
  - Fusion of hydrogen to helium requires temperatures greater than 10,000,000 Kelvin.
  - Fusion of helium to carbon requires temperatures greater than 100,000,000 Kelvin.

- If gravity in a region of a star is **greater than (>)** the outward pressure (gravity > pressure), that region will contract and therefore heat up (it acts as an ideal gas).

- If the outward pressure is greater than the gravity (pressure > gravity) in a given region of a star, that region will expand and therefore cool (it acts as an ideal gas).

1. There are two statements of "it acts as an ideal gas" in the above list. When a gas contracts and heats, this means that the thermal pressure will

   a. increase. **b.** decrease. **c.** stay the same.

Step 2—The Sun's Post-Main-Sequence Evolution

We can summarize the stages of the evolution of the Sun on the H-R diagram, starting with the Sun on the main sequence (A) and ending with it cooling as a white dwarf (G). The location of the Sun on this diagram gives its approximate temperature, luminosity, and radius at that time in its evolution. The stages of evolution are identified in **Figure 21.1**.

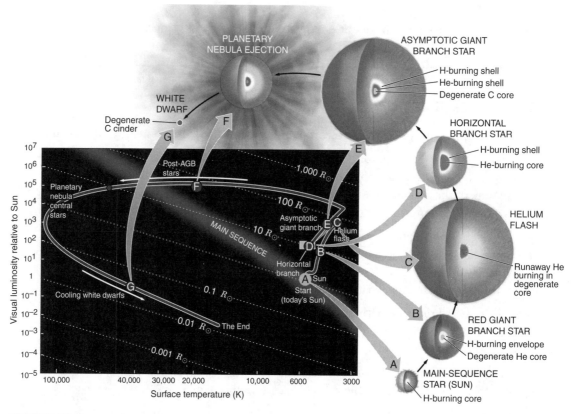

FIGURE 21.1

2. Complete **Table 21.1** by comparing two stages and indicating whether the temperature, luminosity, and radius each increased, decreased, or stayed the same. We have shown an example for the changes from stage A to stage B.

The changes that you listed in Table 21.1 are observable changes that astronomers might detect if they watched the Sun evolve from afar. These observable changes are caused by changes in the amount, location, and type of fusion that is happening within the star.

⊙ TABLE 21.1

Comparing the stages of the evolution of the Sun

CHANGES OCCURRING	TEMPERATURE	LUMINOSITY	RADIUS
from A to B	decreased	increased	increased
from B to C			
from C to D			
from D to E			
from E to F			
from F to G			

Step 3—An Analysis of the Seven Identified Stages

Combine all of the information from the background, Figure 21.1, and Table 21.1, and fill in Table 21.2 as follows:

- For columns 2–5, determine from Figure 21.1 where fusion is occurring—in the core or in the shell, or both—and the kind of fusion process taking place.
- For column 6, check when electron degeneracy is present in the core.
- For columns 7–9, indicate what is supporting the rest of the star.
  - If the star is in equilibrium, then pressure and gravity are in balance.
  - Put a check in column 8 if the rest of the star is contracting (gravity > pressure).
  - Put a check in column 9 if the rest of the star is expanding (pressure > gravity).

● TABLE 21.2

Evolution of the Sun—a step-by-step analysis

| | FUSION | | | | PRESSURE VERSUS GRAVITY | | | |
| | If fusion is occurring in core or in shell around the core, mark with a ✓ in the correct column. | | | | CORE | REST OF THE STAR | | |
Status of Sun's core and its temperature (The letters correspond to the stages in Figure 21.1.)	Core fusion H to He	Core fusion He to C	Shell fusion H to He	Shell fusion He to C	Pressure comes from electron degeneracy	Pressure = Gravity	Gravity > Pressure region contracts and heats	Pressure > Gravity region expands and cools
3. Stage A: T > 10 million K. Sun is in equilibrium.	✓					✓		
4. Stage B: T < 100 million K. Core collapse stops.								
5. Stage C: Helium flash								
6. Stage D: T > 100 million K. Sun is in equilibrium.								
7. Stage E: T < 800 million K. Core collapse stops.								

8. Stage F:

   a. What happens to the rest of the star when stage F on the H-R diagram is reached?

   b. What is this object called, and what is the fate of the material?

After Stage F and before "The End" on the H-R diagram, stage G is reached by the Sun.

9. Stage G:

   a. What is this object called?

   b. Compare the size of the Sun relative to what it was when on the main sequence.

Step 4—Putting It Together

10. Using the key terms, write a summary that describes the future of the Sun in terms of this "battle" between the pressure created by fusion in the Sun (either in the core or in a shell) and the force of gravity trying to squeeze it into a tiny sphere. Fusion converts mass to energy in the form of very high-energy light, known as gamma ray radiation. Where do the high temperatures in the core come from? What will the Sun's final composition be? Consider including the size of the Sun at each stage. If we were able to view it from afar, how much would we see it expand and shrink?

Name _____ Date _____ Section_____

● ACTIVITY 22
The Stuff between the Stars

Learning Goals

In this activity, you will study images of various locations in the interstellar medium to learn to distinguish among the reasons an object might look red. You will also learn to

1. distinguish among the red emission radiation created in H II regions, interstellar reddening caused by dust, and dust clouds emitting infrared light due to thermal radiation.

2. explain how astronomers help us to "see" objects and features that give off light that is outside of the visible part of the spectrum.

Key terms: interstellar medium, nebula, interstellar reddening, thermal emission, H II regions, emission nebula, dark nebula

Step 1—Background

The space between the stars is not empty. It is filled with gas and dust, called the **interstellar medium,** which in some places is quite dense. **Nebulae** are clouds of dust and gas that are dense enough to be distinguished from the surrounding material. This material, even when it is quite diffuse, scatters blue light (with wavelengths between 450–495 nm) in all directions. Stars seen through the interstellar medium appear redder than they actually are because the blue light has been scattered away from the line of sight, leaving the red light (with longer wavelengths around 620–750 nm) behind. This effect is called **interstellar reddening** and can cause stars to look red for a completely different reason than the **thermal emission** that they radiate because of their temperature. This is only one effect of the interstellar medium.

The interstellar medium can be observed directly as well. **H II regions** are regions of space in which the hydrogen is mostly ionized by nearby hot stars. These regions glow because they are hot. A similar effect can occur in any region of the interstellar medium, with electrons in various types of elements jumping up and down the energy levels and creating emission lines. A nebula that glows for this reason is called an **emission nebula.** Sometimes the dust and gas are too thick to glow in the visible part of the spectrum, and the nebula traps the visible light inside. These **dark nebulae** appear dark brown or black in images and are usually seen in silhouette against a background of brighter emission nebulae. Sometimes these nebulae are bright in the infrared, while being dark in the visible.

1. An image taken in visible light includes wavelengths between _____ nm and _____ nm (approximately).

2. The wavelength range for the red part of the spectrum is roughly between _____ nm and _____ nm.

Step 2—Mapping the North America Nebula

Review the two images of the North America Nebula, shown in **Figure 22.1** in the appendix: one is an image of the nebula taken in visible light, and the other is an image of the nebula taken in infrared by the Spitzer Space Telescope.

3. Start by first making a map of the North America Nebula imaged at visible wavelengths in **Figure 22.2**. Fill in just enough of the details so that you can distinguish the emission nebula (reddish) from the dark nebula (dark gray to black). Indicate the positions of a few of the brighter stars. The Pelican Nebula from the right-hand side of the visible wavelength image has already been sketched in Figure 22.2 as an example of mapping. Be sure to label or provide a key for your part of this sketch.

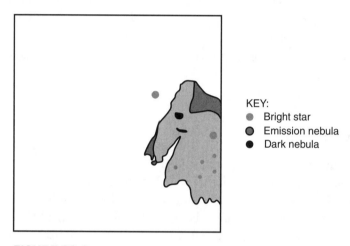

FIGURE 22.2

4. Now let's do a more detailed comparison of the visible and infrared images. List two significant ways that the visible-light image differs from the infrared-light image.

5. What mechanisms, events, or processes are producing the light in each image?

Step 3—Discovering Why Interstellar Reddening of Light Occurs

Study the color image of the Horsehead Nebula, shown in **Figure 22.3** in the appendix. Studying this object will guide you to a better understanding of the reddening of starlight due to interstellar dust. **Figure 22.4a** is a grayscale image of the top region of the Horsehead Nebula in the constellation of Orion. **Figure 22.4b** shows what the Horsehead would look like if you could view it from the side. (Imagine "turning" the image counterclockwise and looking at it from that perspective.) The light from the stars indicated by arrows is being "de-blued"; that is, the blue light from the stars is scattered away from our line of sight, leaving the longer red wavelengths to be seen. This redness results from an entirely different process than that of a star whose blackbody radiation peaks at red wavelengths.

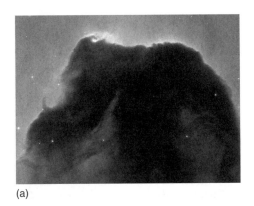
(a)

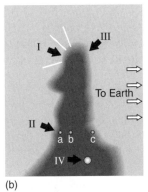
(b)

FIGURE 22.4

6. Which stars in the color image seem to be "reddened" by the dust of the Horsehead Nebula? Circle those stars in Figure 22.4a.

7. In Figure 22.4b, where would a likely location be for those stars whose light has been reddened (locations are labeled, a, b, and c in the figure)?

   a. behind the dust column   **b.** deep inside the dust column

   c. in front of the dust column   **d.** either a or b are possible locations for reddening to occur

8. Indicate on Figure 22.4b where the scattered blue light would go that caused the star's light to be reddened.

9. Is this reddened light coming from the stars seen at visible or infrared wavelengths? Explain.

10. In the sketched "side view" shown in Figure 22.4b, where would you locate the "young star still embedded in its nursery of gas and dust"? _____ I, _____ II, _____ III, or _____ IV?

Step 4—Infrared Light from Warm Dust

An image from the Spitzer Space Telescope titled "Stars Brewing in Cygnus X" is shown in **Figure 22.5** in the appendix. This image was chosen because it encompasses an angular size of the night sky that is around eight times the size of the full Moon yet is invisible to the naked eye.

11. From the Spitzer image, blue represents a mapping of the warmest dust. The wavelength being mapped is 3600 nm. What is the temperature of that dust?

_____

12. The color red represents a mapping of the coolest dust. The wavelength being mapped is 24,000 nm. What is the temperature of that dust?

_____

13. The image of this region of space displays the colors blue, cyan, green, and red. Does this mean that the nebula is emitting light in these colors? Explain.

14. Do the red-colored parts of this region indicate emission from hydrogen atoms, reddening of starlight, or dust glowing because it is warm? Explain.

Step 5—Putting It Together

15. Using at least three key terms of this activity, describe emission from hydrogen atoms, interstellar reddening, and thermal emission from dust. Give an example of each process.

16. How do astronomers help us to "see" things that give off light at infrared wavelengths?

○ ACTIVITY 23
Investigating the Crab Nebula and Pulsar

Learning Goals

In this activity, you will investigate a supernova remnant, finding the speed at which a feature embedded in this explosion is traveling outward. You will also

1. confirm which images had to be taken from a space-based observatory.

2. hypothesize as to why the images taken at different wavelengths have different angular sizes (images have the same scale).

3. calculate the speed of a distinctive "wave" from the pulsar and compare it to the speed of light.

Key terms: neutron star, black hole, supernova remnant, pulsar, gamma ray, radio wave, synchrotron radiation, X-ray, emission line, radio, infrared, shock wave

Step 1—Background

When massive stars die, they leave behind two distinct objects: the dense central object that is either a **neutron star** or a **black hole**, and an expanding cloud of gas called a **supernova remnant**. In either case, the central object will have extremely strong gravity. Neutron stars also have a strong magnetic field. Rotating neutron stars can produce **pulsars** when the magnetic field axis points directly at Earth repeatedly as the neutron star spins. The pulsing light provides a way to detect these very dense stellar corpses.

The supernova remnant ejected from the dying star expands rapidly, cooling down as it moves farther from the center of the explosion. Supernova remnants are often observable at all wavelengths across the electromagnetic spectrum, from **gamma rays** to **radio waves**. Observations in different parts of the electromagnetic spectrum each show something different about the supernova remnant. Scientists combine these observations together to get a more complete picture of what is happening in the expanding gas than can be seen in any single observation. In the innermost regions, the light from the supernova remnant is dominated by **synchrotron radiation**. This is light emitted by electrons circling very, very fast around magnetic field lines; it is emitted in the **X-ray** region of the spectrum. Much of the visible light from the supernova remnant comes from electrons reconnecting with atomic nuclei to produce **emission lines**. Light in the **radio** and **infrared** regions of the electromagnetic spectrum is emitted from the cooling gas as the supernova remnant expands.

As the supernova remnant expands, its structure appears to change, with parts of the nebula appearing, disappearing, and/or moving. Figuring out precisely what is happening can be difficult, because there are three different mechanisms for making a part of the nebula bright. First, there may be a lot of material at that spot; second, there might be a lot of light there (coming from the central object, for example, and ionizing the material in that part of the nebula); third, there may be a **shock wave** moving through the nebula. A shock wave is a pressure wave (like sound) that can travel through a medium very fast, heating up the material as it passes by. Because these three factors are in play, if you see a bright spot appear to move from one place to another in a supernova remnant, it might be material moving, or it might be caused by a moving "searchlight" from the central object, or it might be a shock wave passing through. The material itself might not have actually moved from one place to another. Often, calculating the speed of the apparent motion can help distinguish among these possibilities.

The Crab Nebula is a supernova remnant from a supernova that was observed in A.D. 1054. This supernova remnant has been expanding outward into space since then. In the center of the Crab Nebula is a rapidly rotating neutron star that has a magnetic field axis that sweeps past Earth; it is a pulsar that pulses 60 times each second.

Recall that, like visible light, radio waves pass easily through Earth's atmosphere. All other regions of the spectrum are blocked by the atmosphere. To observe these regions of the spectrum astronomers must use telescopes placed above most of the atmosphere, either in space or on the top of a very high mountain.

1. All pulsars are neutron stars, but not all neutron stars are pulsars. What makes a neutron star a pulsar?

Step 2—The Crab Nebula

The Crab Nebula is shown in **Figure 23.1** in four wavelength regions. Consider these images, and then answer the following questions.

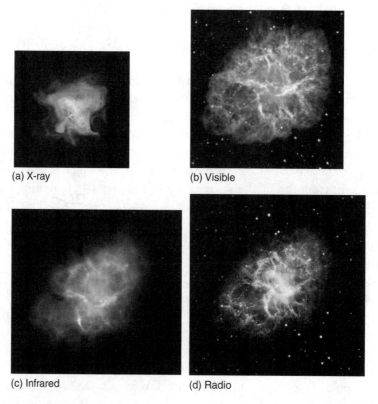

(a) X-ray (b) Visible

(c) Infrared (d) Radio

FIGURE 23.1

2. Think about the regions of the electromagnetic spectrum where light reaches Earth's surface and other regions where the atmosphere blocks the light. Ignoring any sharpness or clarity of the images shown here, which two of the images *had* to be taken either from a high mountain or a space telescope?

3. The X-ray image shows where high-energy events are happening, while the radio region shows where low-energy events are happening. Approximately how much larger is the radio-emitting part of the supernova remnant than the X-ray-emitting part?

4. Based on your knowledge of synchrotron radiation (**hint:** see Background section), does the difference in size of these two regions make sense? Explain.

Step 3—The Crab Pulsar

Recent observations of the Crab Nebula have detected rapidly changing wisps of material. These changes are shown in the Hubble Space Telescope images shown in **Figure 23.2**. These include wisplike structures that move outward away from the pulsar at almost half the speed of light, as well as a mysterious "halo" that remains stationary, but changes in brightness over time. **Figure 23.3** "zooms in" to the core of the nebula: the region around the pulsar. Based on the original image scale, it appears that the white line, depicting a "wave," moved about 6,000 astronomical units (AU) in just 75 days between February 1 and April 16, 1996.

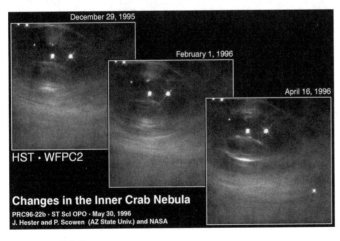

FIGURE 23.2

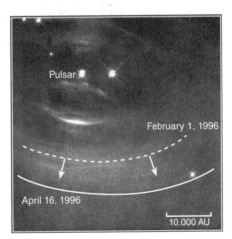

FIGURE 23.3

5. What is the approximate speed for the white line in kilometers per second (km/s)? You need the following conversions to calculate the answer to this question: 1 AU = 150,000,000 km; 1 day = 60 × 60 × 24 = 86,400 seconds.

   Setup: (6,000 AU × 150,000,000 km/AU) / (75 days × 84,400 s/day) = speed

   Speed of white line: _____ km/s

6. At what fraction of the speed of light, 300,000 km/s, was this wave moving? (Show the equation and the numbers you used.)

   Speed as fraction of speed of light: _____

7. Consider your calculation of the speed of the material. Is this a reasonable speed for the material to travel? Could it be a shock wave that is fooling us? Explain your thoughts.

8. The diameter of a neutron star is approximately 20 km. How much of your city would the neutron star cover if it were to sit on a pedestal, say at city hall? Bring in surrounding neighborhoods, if necessary, and explain the numbers you use.

Step 4—Putting It Together

9. Using key terms of this activity, explain how the Crab Nebula and pulsar formed. Compare the four images in Figure 23.1 and the images shown in Figure 23.2 to support your explanation.

● ACTIVITY 24
The Life and Death of Massive Stars

Learning Goals

After reviewing the events taking place during the life and death of a massive star, you will be able to

1. correctly order the sequence of core fusion cycles.

2. identify the stages where overlapping or simultaneous events occur.

3. determine if a neutron star or a black hole will be the end result of evolution for a particular star.

Key terms: CNO cycle, catalyst, degenerate, reaction cycle, fusion shell, neutron degeneracy, neutron star, pulsar

Step 1—Background

Stars with masses greater than about 8 solar masses ($8\ M_{Sun}$) are located at the top of the H-R diagram, with luminosities many thousands or even millions times that of the Sun. Their surface temperatures vary between around 40,000 K and 4000 K as they evolve. Even though these stars are massive, their high luminosities mean their lifetimes are much shorter than that of the Sun, living only millions of years compared with the Sun's many billions of years.

In the interior of a massive star, there is an extreme, ongoing battle between the inward pull of gravity and the outward push of pressure from radiation and from the kinetic energy of particles. Nuclear fusion in the core of these stars ultimately produces this pressure, and in doing so, must fuse heavier and heavier elements. Once iron is reached, the star can no longer support itself against the inward pull of gravity.

Because the core of a massive star has much higher temperatures and pressures than that of a Sun-like star, when it is on the main sequence it can add the **CNO cycle** (carbon-nitrogen-oxygen) for fusion of hydrogen to helium. In this process, carbon nuclei act as a **catalyst**, which helps the reactions occur, but is not used up in the fusion reactions. Another distinction from the evolution of the Sun is that the core of a massive star does not become **degenerate** between cycles (where the electrons are pushed as close together as rules of quantum mechanics allow); rather, the core proceeds from one **reaction cycle** to the next, becoming much hotter, smaller, and denser as it does so. Each cycle leaves behind a **fusion shell**. A massive star may end up with many of these shells as it nears the end of its life.

When a massive star dies, it does so spectacularly, generating a blast that creates a luminosity equal to that of an entire galaxy. This explosion is called a Type II supernova. What is left depends on the mass that remains.

A star with an initial mass of more than $8\ M_{Sun}$ but less than about $25\ M_{Sun}$ leaves a core with mass less than about $3\ M_{Sun}$. For these cores, the collapse is halted when neutrons are packed as tightly together as the rules of quantum mechanics allow; the material of the star is **neutron degenerate** matter. This object is a **neutron star**, and it is very small for its mass. As the core of a massive star, its radius may have been 0.1–0.01 times the radius of the Sun. As a neutron star, its radius is about the size of a large city: 10 km.

If the neutron star spins rapidly and has a strong magnetic field, it could be a **pulsar**. As the star shrank, angular momentum was conserved, so the neutron star spins faster than the original star. Newly formed pulsars rotate hundreds of times each second. The magnetic field the massive star had originally is compressed and becomes much stronger in the neutron star. Our neodymium magnets are very strong; a quarter-sized magnet can lift 9 kg (20 pounds). The magnetic field of a neutron star is a million times stronger. If the axis of the magnetic field does not line up with the rotation axis, then the magnetic poles will sometimes point toward Earth, and sometimes away, causing the star to appear to blink on and off like a lighthouse. From our point of view, all pulsars are neutron stars, but not all neutron stars are pulsars, because some have their magnetic poles pointing in different directions.

Stars with initial masses more than about 25 M_{Sun} will leave cores that are more than 3 M_{Sun}. If this occurs, then gravity will be stronger than the pressure provided by neutron degeneracy, meaning nothing can stop the collapse. A black hole is born: an object with gravity so strong that not even light can escape.

1. The evolution of stars having masses greater than about 8 M_{Sun} is different from the evolution of the Sun because

   a. their cores do not experience electron degeneracy between fusion cycles.

   b. fusion of elements much heavier than helium occurs.

   c. they end up as a neutron star or a black hole instead of a white dwarf.

   d. All of these answers describe the differences between the masses.

Step 2—Visualizing the Evolution of a Massive Star

Although we do not know exactly what occurs deep inside a star more massive than the Sun, we have models that give us a pretty good idea. It helps to visualize the life of one of these stars to understand the stages they go through. We start with a star at the top of the main sequence, as shown in **Figure 24.1**. Read through the details of each of the four identified stages of a massive main-sequence star, and then move to Step 3.

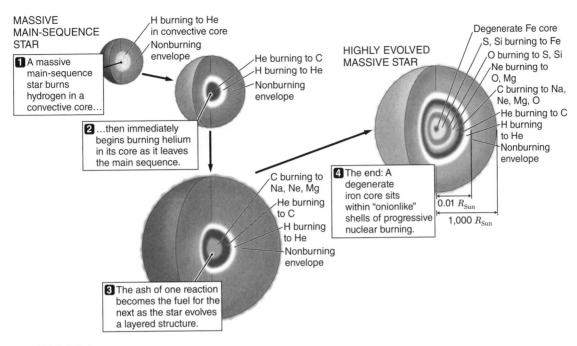

FIGURE 24.1

Step 3—Understanding the Flowchart of the Evolution of a Massive Star

Figure 24.2 shows a flowchart that starts with the massive star on the main sequence and proceeds through its becoming a Type II supernova. Each stage—lettered A to H—is represented by a cell in the flowchart that contains one or more questions for you to answer. Use the information given in Steps 1 and 2 and answer the questions that follow.

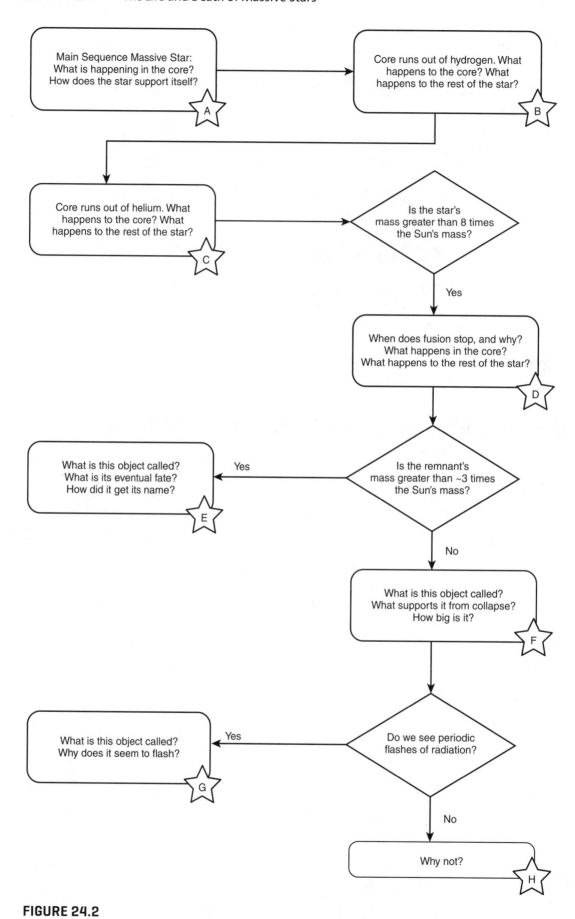

FIGURE 24.2

2. Cell A: As a main-sequence massive star, what is happening in its core? How does the star support itself?

3. Cell B: When the core runs out of hydrogen in a few million years, what happens to it? What happens to the rest of the star?

4. Cell C: Time passes, and the core runs out of helium. What happens to the core? What happens to the rest of the star?

5. Cell D: We've ascertained that we are dealing with a massive star. When does fusion stop, and why? What happens in the core? What happens to the rest of the star?

6. Cell E: We have determined that the mass of the remnant is much greater than can be supported by any kind of pressure, including neutron degeneracy. What is the object that results called? What is its eventual fate? How did it get its name?

7. Cell F: The mass of the remnant on this branch is less than about three times the mass of the Sun. What is this object called? What supports it from collapse? How big is it?

8. Cell G: We observe the remnant on this branch emitting periodic flashes of light. What is this object called? Why does it seem to flash?

9. Cell H: We know the kind of remnant we are observing, but we do not see periodic flashes of radiation. Why not?

Step 4—Putting It Together

10. In a few paragraphs, highlight the stages that a massive star passes through as it evolves from a main-sequence star to its dramatic end. You may choose to compare and contrast its evolution to that of the Sun or emphasize one stage of the massive star evolution over another. You should be able to use at least half of the key terms in this activity.

● ACTIVITY 25
Determining Ages of Star Clusters

Learning Goals

In this activity, you will find the ages of star clusters (and the stars in them) by considering the rate at which stars fuse hydrogen to helium and by studying the color-magnitude diagrams of clusters. After completing the activity, you will be able to

1. state the relationship between the color of a main-sequence star and its main-sequence lifetime.

2. list the steps for determining the age of a cluster given the cluster's "turnoff" B–V color.

3. determine cluster ages using their color-magnitude diagrams.

4. summarize the complete process of how astronomers estimate the ages of star clusters.

Key terms: magnitude, apparent magnitude, luminosity, B–V color index, color-magnitude diagram, star cluster, main sequence, nuclear fusion, fuse, spectral type, turnoff

Step 1—Background

The brightness of a star at visual wavelengths is typically measured in **magnitudes**. The magnitude scale goes far back into astronomical history and has some odd characteristics. First, the scale is logarithmic so that a difference of one magnitude corresponds to a factor of about 2.5 in brightness. Second, the scale runs backward. When the scale was developed, the brightest stars were called magnitude 1, the next brightest were magnitude 2, and so on, down to the faintest stars, which were magnitude 6. While this is a perfectly logical way to rank stars by brightness, it means that a brighter star has a smaller magnitude. The **apparent magnitude** of a star (usually just called "magnitude") measures how bright the star appears in our sky. This is different from the **luminosity** of the star, which measures how much light the star emits.

Modern astronomers measure the brightness of stars through special filters placed on the detectors they use on their telescopes. What they find is that a star might have a B (blue) magnitude and a V (visual; yellow-green) magnitude, which are not the same. Hot stars are bluer than cool stars. Conversely, cool stars are redder than hot stars. The **B–V color index** (often shortened to just "B–V") quantifies this observation using two different filters. To find the B–V color index, astronomers first measure the apparent magnitude with two different filters: B and V. They then subtract the V magnitude from the B magnitude to find the color index. Theoretical spectra for objects at two different temperatures and filter wavelengths for B and V filters are shown in **Figure 25.1**. The hottest stars have B–V color indices close to –0.5 or less, while the coolest stars have B–V color indices close to 2.0. Notice that bluer, hotter stars have LOWER B–V indices, because the magnitude scale is backward. Other stars are somewhere in between.

A **color-magnitude diagram** plots the V magnitude versus the B–V color index. A color-magnitude diagram is another way of making an H-R diagram. Instead of plotting the luminosities of the stars and their temperatures, we use their apparent magnitudes and their colors. We can do this for **star clusters** because the stars in them are all at basically the same distance, and we know how temperature relates to their colors.

As you likely know from experience, it is possible for a car to hold more fuel but run out of that fuel faster than a car that holds less fuel. This idea applies directly to stars. Stars spend most of their lives on the **main sequence**, which extends from lower right to upper left on a color-magnitude diagram. Stars on the main sequence convert the hydrogen in their cores into helium and energy through **nuclear fusion**. Stars that are more massive and have more hydrogen in their cores will burn (**fuse**) that fuel faster, be hotter, and always run out of hydrogen fuel sooner than those that are less massive. Put another way: The more massive a star is, the hotter it is, and the shorter its life span.

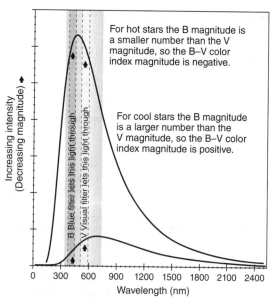

FIGURE 25.1

1. You observe two main-sequence stars, star A and star B. Star A is bluer than star B.

   a. Which star has the smaller B–V color index?

   b. Which star is hotter?

   c. Which star has a shorter life span?

Step 2—B–V Color Index and Main-Sequence Lifetime

2. Study the information about the three main-sequence stars shown in **Table 25.1**. The second column indicates the **spectral type** of these stars in order of temperature, with spectral types given the letters OBAFGKM. The hottest stars are O stars, and the coolest ones are M stars. Which star(s) is/are brighter in the B filter than in the V filter?

⚫ TABLE 25.1

B and V color indices for three main-sequence stars

STAR	TYPE	B-FILTER MAGNITUDE	V-FILTER MAGNITUDE	B–V COLOR INDEX	HOT, HOTTER, HOTTEST?
Barnard's Star	M4 V	11.24	9.51	1.73	hot
Mintaka	O9 V	2.01	2.41	−0.4	Hottest
Zavijava	F9 V	4.16	3.61	0.55	Hotter

Learning Astronomy by Doing Astronomy Second Edition

3. Calculate the B–V color index for the stars by subtracting the V magnitude from the B magnitude, and then fill in the fifth column of Table 25.1.

4. Rank the three stars in order of temperature and record the ranking in the last column.

5. Which of these three stars will spend the shortest amount of time on the main sequence?

6. Which will spend the longest amount of time on the main sequence?

7. These answers imply that there is a relationship between a star's B–V color index and its main-sequence lifetime. What is that relationship?

Step 3—Quantifying the Relationships

So far, we have been somewhat *qualitative* in our descriptions of how B–V index is related to main-sequence lifetime. There is a *quantitative* relationship between these two quantities that we now consider. We find this relationship by studying star clusters, those large groups of stars that were all born at the same time from the same cloud of dust and gas. In a star cluster, all the stars are the same age as each other AND they are the same age as the cluster. The uppermost point at which the main sequence ends is called the main-sequence **turnoff**. As a cluster ages, the main-sequence turnoff changes in a very predictable way. As time passes, the most massive, hottest, most luminous stars die first, and the main-sequence turnoff moves down and to the right, shortening the main sequence. Star clusters have an additional convenient property: all of the stars in a star cluster are at the same distance from us. This means that the apparent magnitudes measure the relative luminosities among the observed stars. This is not true in a random star field, in which a star might be faint because it has low luminosity, or it might be faint because it is farther away than other stars.

8. Study the data in **Table 25.2**. How does the age of a star cluster depend on the value of the B–V color index of the turnoff point from the main sequence?

O TABLE 25.2

Comparison of stellar properties among spectral types

PARAMETER	SPECTRAL TYPE						
	O5	B0	A0	F0	G0	K0	M0
Mass (solar masses)	40	15	3.5	1.7	1.1	0.8	0.5
B–V color	**−1.2**	**−0.3**	**0.0**	**0.3**	**0.6**	**0.8**	**1.4**
Main-sequence lifetime (years)	1.0×10^6	1.1×10^7	4.4×10^8	3.0×10^9	8.0×10^9	1.7×10^{10}	5.6×10^{10}

Overlapping color-magnitude diagrams for two different star clusters, NGC 2362 and M68, are shown in **Figure 25.2**. Note the numbered star symbols for NGC 2362 stars and numbered solid circles for M68 stars.

Stars in the cluster NGC 2362 are marked with star symbols. The turnoff for NGC 2362 appears to be at a B–V value of 0.0. Stars in the cluster M68 are marked with solid circles. The main-sequence turn-off for M68 appears to be at a B–V value of about 0.4.

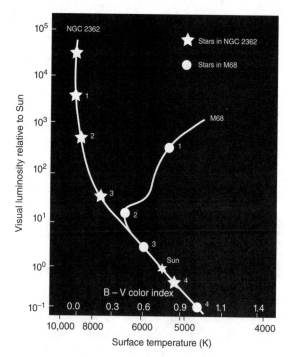

FIGURE 25.2

9. Use Table 25.2 to answer the following questions:

   a. What is the approximate age of the cluster NGC 2362?

   _____

   b. What is the approximate age of the cluster M68? _____

10. Which of the following statements is correct?

    a. Stars in M68 are older than stars in NGC 2362.

    b. Stars in NGC 2362 are older than stars in M68.

    c. The stars in each cluster have a large range of ages.

    d. The stars in M68 are the same age as the stars in NGC 2362.

Step 4—Applying the Relationship to Real Star Clusters

Imagine that you observed four different clusters in our galaxy, and created a color-magnitude diagram of each one, as shown in **Figure 25.3**. The main sequence of each cluster clearly extends upward toward the upper left from the lower right.

11. Rank the ages of the different clusters shown, from youngest to oldest.

    Youngest cluster _____ _____ _____ _____ Oldest cluster

    Explain your logic.

Learning Astronomy by Doing Astronomy Second Edition

In astronomy, we rarely get to watch a single object evolve over time, because it takes too long. Instead, we observe many objects at different stages, and put those stages together to determine how a single object would change over time. Imagine that the color-magnitude diagrams in Figure 25.3 show a snapshot of the stars in a *single* cluster, with each panel taken at a different time throughout the history of the cluster.

12. Rank the plots in the order in which they would occur.

First _____ _____ _____ _____ Last

Explain your logic.

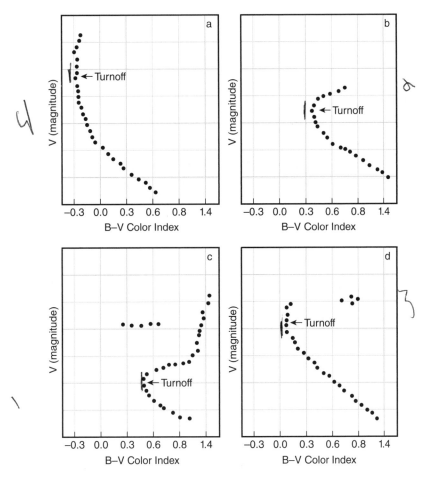

FIGURE 25.3

Real clusters present a more difficult challenge. Creating a color-magnitude diagram means taking two images of the cluster, one in each filter. These images contain stars that are not part of the cluster, but instead are in front of it, or behind it. This adds an element of noise to the data, so that the main sequence is less well-defined. It is typically still visible in a color-magnitude diagram, though, as the densest grouping of stars on the graph.

13. Determine the B–V value of the turnoff of each of the clusters in **Figure 25.4** by reading down to the *x*-axis from the turnoff point, and then use Table 25.2 to determine the age. Place your results in **Table 25.3**.

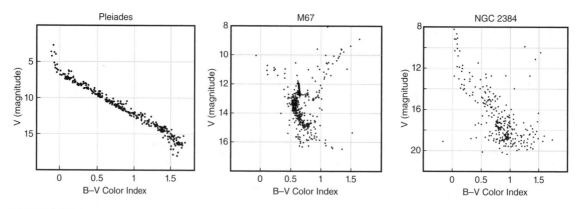

FIGURE 25.4

O TABLE 25.3

B and V color indices of the turnoff and age for three star clusters

CLUSTER	B–V OF TURNOFF	AGE
Pleiades		
M67	0.6	8.0×10^9
NGC 2384		

Step 5—Putting It Together

14. Explain how a color-magnitude diagram allows us to determine the age of a cluster. This should *not* be a step-by-step how-to; rather, your response should incorporate ideas and key terms covered in this activity and explain how at least three of the following topics are involved: how clusters form; distance to a cluster; luminosity of the stars; lifetimes of stars based on their masses; and what astronomers mean by the turnoff of the main sequence.

● ACTIVITY 26
Bent Space and Black Holes

Learning Goals

In this activity, you will use thought experiments and the metaphor of a rubber sheet to explore the interaction of black holes with objects that pass near them. As you work through this activity, you will also

1. explain the concept of bent space.

2. demonstrate an understanding of how the speed of an object affects its interaction with a massive body.

3. define *event horizon.*

4. extend the concept of an event horizon to objects that travel at speeds slower than light.

5. extrapolate the two-dimensional concepts to three dimensions.

Key terms: black hole, event horizon, trajectory

Step 1—Background

Einstein's idea of objects bending or warping space applies to anything having mass—the more massive an object is, the more space is bent. Here we will consider the ultimate in bent space, that caused by **black holes**, which bend space so severely that objects would have to travel faster than light to escape from the "pit" in spacetime. Objects far from a black hole experience gravity normally, as though the black hole were a star with the same mass. As objects come closer to the black hole, they have to travel faster to get back out of the warped spacetime again. At a particular distance from the black hole, an object would have to travel faster than light to escape. The sphere around the black hole, located at this distance, is called the **event horizon**. No object can escape the black hole once it has crossed the event horizon, because no object can travel faster than light.

Because we can't actually grab a black hole and bring it into the lab and because scientists have never actually observed one directly, we can conduct only "thought experiments" to explore their properties. These thought experiments involve a big rubber sheet, like a trampoline, that is very stiff and not easily stretched, but does have some "give" to it. This sheet is held parallel to the ground. Adding a heavy object, like a bowling ball, to the sheet will make it bend downward. Smaller objects, like golf balls, are light enough to make no perceptible bend in the sheet. A bend in the sheet changes the **trajectory**, or path, of objects that roll across it, just as the presence of mass in space changes the trajectory of objects that move past it.

1. Suppose the Sun were replaced by a black hole with a mass equal to the mass of the Sun. The event horizon of such a black hole is about 6 km across. Would Earth orbit normally, or would it spiral into the black hole? Explain.

Step 2—Black Holes in Two Dimensions

2. Imagine a big rubber sheet, as shown in **Figure 26.1**. Imagine that you roll some golf balls across it in directions you choose. Sketch their paths across the sheet on the figure.

3. Now imagine that we put a bowling ball in the middle of the sheet to make a big, slope-sided pit, as shown in **Figure 26.2a**. (A perspective view is provided in **Figure 26.2b** to aid in your visualization.) Sketch the paths of three golf balls rolled across the sheet on the following trajectories:

   a. The golf ball is far from the bowling ball, near the edge of the sheet.

   b. The golf ball goes directly toward the bowling ball.

   c. The golf ball comes close to the bowling ball but not directly at it.

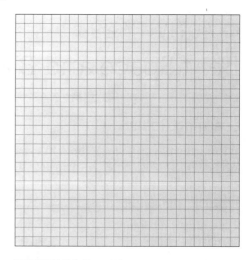

FIGURE 26.1

4. In each of the three cases of question 3, how would things change if the golf balls were moving very, very quickly?

5. In each of the three cases of question 3, how would things change if the golf balls were moving very, very slowly?

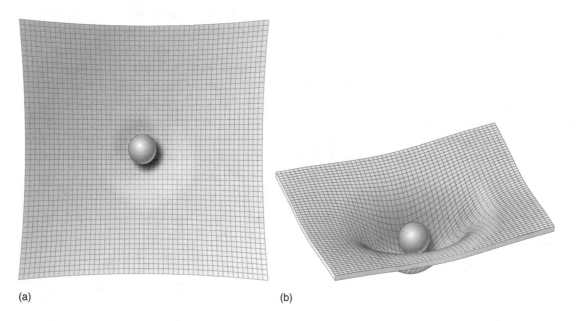

(a)   (b)

FIGURE 26.2

6. Imagine that you send lots of golf balls into the pit. What happens to the depth and width of the pit as the golf balls fall into the center near the bowling ball?

7. So far you have been working with two examples of space. In the first example, the space is flat; in the second example, a massive object placed in the center bends space quite a bit. How do these examples explain the concept of bent space in terms of the objects rolling across the rubber sheet?

The pit is a fair analogy for a black hole. Objects outside the pit will "know" that the black hole is there because the sheet is sloping, but they won't be captured unless they come within the event horizon. This is the location at which light bends exactly into a circle around the black hole, as illustrated in **Figure 26.3**, which shows the event horizon and black hole in perspective, as in Figure 26.2b. Think about light as though it were grains of sand rolling across the sheet extremely fast.

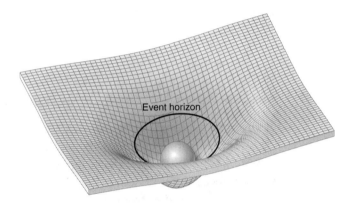

FIGURE 26.3

8. Sketch the path of one grain-of-sand light particle if it were to travel

   a. far from the bowling ball, near the edge of the sheet.

   b. directly toward the bowling ball.

   c. on a line that is tangent to (just touches) the event horizon.

   d. inside the event horizon, but not directly at the bowling ball.

Stars, people, planets, all things interact in this way because of gravity. In the case of black holes, it is a little more accurate to think of the bowling ball as making a very, very deep hole (so nothing can bounce out, for example), but the sheet is still bent in the same way.

9. Now think about what happens when you roll the slower-than-light golf balls past the pit. Sketch and label a circle on Figure 26.3 that would mark a "golf ball event horizon." If the golf balls come closer to the hole than this circle, they cannot escape.

10. Is the "golf ball event horizon" that you sketched closer or farther from the bowling ball than the light event horizon? Why? (**Hint:** Think about how the speed affects the motion of objects around the pit.)

11. Suppose that you roll another bowling ball across the sheet. What happens to the sheet when the second bowling ball falls in after the first?

12. How would this affect the paths of the golf balls? How about the example of a grain-of-sand light particle?

13. From your experience with these thought experiments, give a definition of event horizon that is consistent both with golf balls rolling on a rubber sheet that is bent and with light interacting with the bending of space around a black hole.

Step 3—Black Holes in Three Dimensions

The thought experiments that you have just carried out are two-dimensional metaphors for a three-dimensional bending of space. To think about black holes more accurately, you need to add a third dimension in which the same effects apply. This is much more difficult than it might seem, because now the rubber sheet is no longer flat with a pit in the center but is bent from all directions toward the center no matter from which direction golf balls or light approaches.

14. Imagine that a golf ball is fired into Figure 26.3 from above the page.

   a. If the golf ball starts out aimed far from the black hole, at the very edge of the sheet, what will happen to it? Why?

   b. If the golf ball is aimed at a point closer to the black hole, but not within the "golf ball event horizon," what will happen to it? Why?

   c. If the golf ball is aimed directly at the bowling ball, what will happen to it? Why?

Step 4—Putting It Together

15. Using the key terms of the activity, summarize how objects behave near a black hole. How does the mass, velocity, or trajectory of the object affect its behavior? Describe how the rubber-sheet analogy works for understanding the bending of space and how it fails.

● ACTIVITY 27
Dark Matter

Learning Goals

In this activity, you will analyze the rotation of the disk of a spiral galaxy to discover that the luminous mass and the gravitational mass are different. By the end of this activity, you will be able to

1. explain the difference between luminous mass and gravitational mass.

2. estimate the gravitational mass from a galaxy's rotation curve.

3. estimate the luminous mass from a graph of the light interior to a radius.

4. compare these two ways of calculating the mass of a galaxy to rediscover dark matter.

5. summarize how astronomers discover dark matter in a galaxy.

Key terms: dark matter, luminous mass, gravitational mass, rotation curve

Step 1—Background

Dark matter is matter that neither absorbs nor radiates light. We can detect it using gravitational methods (Newton's laws), but not through any interaction with light. This is similar to detecting the wind: you can determine if wind is present through its interaction with trees or grasses or dust, but you cannot see the wind directly. Astronomers do not yet know the composition of dark matter; it is currently one of the mysteries in astronomy and astrophysics.

One way that we find dark matter is to calculate the mass of a galaxy in two different ways, and then compare these values. The first way to find the mass of a galaxy is to add up all the light that comes from it, and then estimate the number of stars that would produce that much light. The mass found in this way is the **luminous mass**. The second way is to observe how the galaxy rotates using the Doppler shift. A galaxy that rotates quickly must have a lot of mass to keep it from coming apart. A galaxy that rotates slowly has less mass. The mass found in this way is the **gravitational mass**. If these two numbers agree, then the luminous mass and the gravitational mass are the same, and there is no mystery. But if the gravitational mass is larger than the luminous mass, then there must be mass present that we cannot see; there must be dark matter in the galaxy.

To find the luminous mass of a spiral galaxy, we must be able to see the whole disk, so ideally, we would observe the galaxy from directly above the plane of the disk of the galaxy. However, to find the gravitational mass, we must be able to measure the rotation of the disk, so ideally, we would observe the disk from the side. We cannot make both of these ideal measurements for any single galaxy because we can observe it only from our single vantage point here on Earth. Instead, we compromise and choose the galaxies we observe to be only those that we see from an angle about halfway between "face-on" and "edge-on." These galaxies appear as ovals, with a long axis and a short axis. The ratio of the short axis to the long axis can be used to find out how tilted the galaxy is, and then we can calculate how our measurements differ from those we would take if we viewed it from above or from the side.

We find the speed of rotation of the galaxy by making a rotation curve. A **rotation curve** plots the radial velocity (the velocity along the line of sight, as measured by the Doppler effect) versus the radius. One side of the galaxy is moving with a positive velocity (away from us), while the other side has a negative velocity (toward us). These data come from spectral lines that have been either blueshifted or redshifted. This curve is then corrected for the viewing angle.

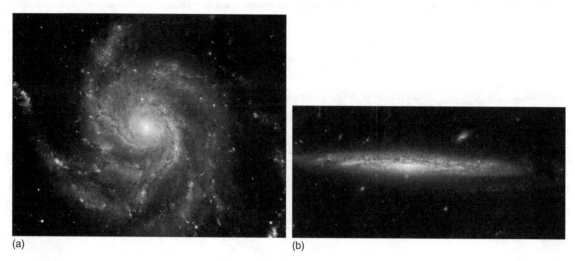

(a)   (b)

FIGURE 27.1

1. The two galaxies shown in **Figure 27.1** cannot be used in a search for dark matter. For each one, explain why not.

   a.

   b.

Step 2—Finding the Gravitational Mass

A rotation curve of NGC 2742 is shown in **Figure 27.2**. This galaxy was chosen because it is neither face-on nor edge-on, but tilted from our point of view. A correction to the measured velocities has been made in Figure 27.2b to account for the tilt of this galaxy. Using Newton's universal law of gravitation, you will be able to determine how much mass the galaxy has based on the force of gravity felt by all of its parts.

2. To see how these calculations are done, use the graph in Figure 27.2b to determine the radial velocity, v, at a radius, r, of 9 kpc from the center of the galaxy.

$$v = \underline{150} \text{ km/s}$$

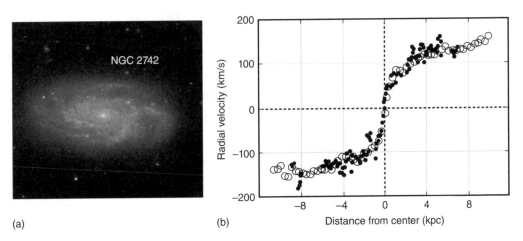

NGC 2742

(a)

(b)

Distance from center (kpc)

FIGURE 27.2

3. Use your value for that radius and radial velocity to determine the mass of the galaxy, M_G, interior to that radius:

$$M_G = 2.3 \times 10^5 \times r \times v^2$$

(The 2.3×10^5 fixes up the units so the answer comes out in units of the mass of the Sun.) Convert your answer to scientific notation.

$$M_G = \underline{4.65 \times 10^{10}} M_{Sun}$$

4. Evaluate your number. Is this a reasonable mass for a section of a galaxy made of billions of stars? Explain.

5. A graph of the gravitational mass versus radius is shown in **Figure 27.3.** Add your data point to the graph. Describe this graph; how does the gravitational mass change as the radius increases?

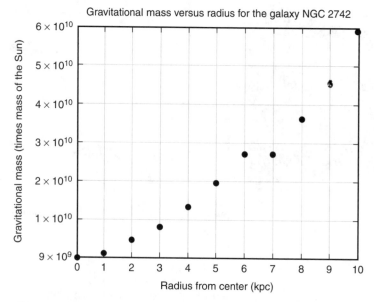

FIGURE 27.3

Step 3—Finding the Luminous Mass

Now that we've found the gravitational mass of the galaxy, we must find out how much of that mass comes from luminous matter like stars and gas. Imagine that we drew a circle around the disk at the radius of 9 kpc. The luminosity within that circular radius is graphed in **Figure 27.4** (along with all the other radii), in units of the luminosity of the Sun.

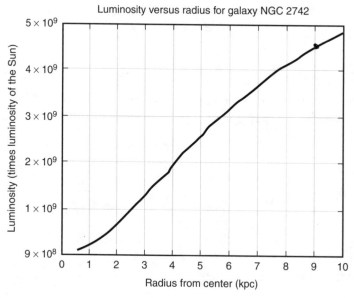

FIGURE 27.4

6. The curve in Figure 27.4 does not drop down at large radii even though the galaxy, like all spiral galaxies, is fainter at large radii. Explain why.

7. Using Figure 27.4, determine the luminosity, L, of the galaxy interior to 9 kpc.

$$L = \underline{4.5 \times 10^9} \, L_{Sun}$$

8. Much of the light may be coming from a small number of young, hot bright stars that are only slightly more massive than the Sun. Many cooler, low-mass stars may be hidden from view. Models of star formation indicate that there are about two solar masses of stars for each solar luminosity of light. Multiply the luminosity from question 7 by two solar masses to estimate how much mass is required to make that much light.

$$M_L = \underline{9 \times 10^9} \, M_{Sun}$$

Step 4—Comparing the Masses

9. In astronomy, we often want an initial estimate of what a graph is "telling" us. Assume that the y-axis gives the luminous mass instead of the luminosity. Compare the values of mass in Figure 27.4 to those in Figure 27.3. Give a general description of how the luminous mass and the gravitational mass differ. What number could you use to multiply each value of the luminous mass at a given radius to approximate the gravitational mass at that same radius?

10. Let's be more exact. Find the percentage of the mass that is luminous by dividing the luminous mass (from question 8), M_L, by the sum of the luminous and the gravitational mass at 9 kpc (from question 3), $M_L + M_G$, and then multiplying by 100 percent to convert the decimal to a percent.

Percent of mass that is luminous: __18%.__

11. Now that you have found the percent of the galaxy's mass that is luminous, the remainder must be dark matter. What percentage of the galaxy's mass is dark matter?

Percent of mass that is dark matter: __82%__

Step 5—Putting It Together

12. Using all of the key terms of this activity, summarize the process used by astronomers to discover dark matter in galaxies, and comment on the amount of dark matter in NGC 2742.

Name _____ Date _____ Section_____

● ACTIVITY 28
Light Travel Time and the Size of a Quasar

Learning Goals

Astronomers use variability in the emission from active galaxies to determine the size of the emitting region and the total energy emitted. In this activity, you will connect these observations to the theory that active galaxies are powered by supermassive black holes. You will also learn to

1. determine the size of the emitting region of a "new" quasar based on a brief intense flare.

2. calculate the energy released by the flare and compare that to the energy of the Sun.

3. evaluate the evidence as to whether it supports the presence of a supermassive black hole.

Key terms: active galactic nucleus (AGN), flare, supermassive black hole

Step 1—Background

An **active galactic nucleus (AGN)** is the region around a supermassive black hole in which material is falling into the black hole. The brightness of an AGN can **flare** or change dramatically very quickly: over the span of months, weeks, days, and even minutes. Astronomers can find the size of an AGN from the speed of light and limits on how quickly an object can change its brightness. These AGN are so powerful that there must be an enormous amount of gravity present to hold them together. They are so small that the mass causing that gravity must be extremely dense. These two observations together lead astronomers to infer that most AGNs host a **supermassive black hole**—a black hole with a mass of hundreds of millions of times the mass of the Sun—in their centers.

Imagine an object 1 light-week in diameter, as sketched in **Figure 28.1**. Suppose that the entire object emits a brief (~1 second) flash of light. Light from the part of the object nearest to Earth will arrive here first. Light from the far side of the object will arrive 1 week later, because this light not only must cross the distance to Earth, but it also must cross the object as well. Because of this delay in the arrival time of the light from the back of the object, we will observe a gradual change in brightness that lasts a full week. Similarly, an object that is 1 light-year in diameter will take 1 year for the brightness to vary.

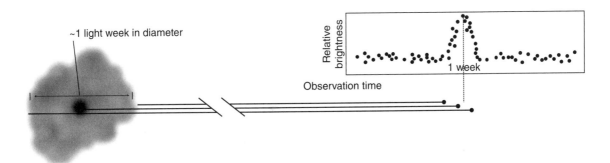

FIGURE 28.1

The size of the emitting region, d, can be found by multiplying the speed of light, c, by the variation time, Δt:

$$d = c \times \Delta t$$

This variation time must be in seconds. For the example of an AGN with a 1-week variation time, we must first convert 1 week (7 days) into seconds:

$$\Delta t = 7 \text{ d} \times 24 \text{ h/d} \times 60 \text{ min/h} \times 60 \text{ s/min}$$

$$= 604{,}800 \text{ s}$$

Then we can multiply this time by the speed of light to find the diameter:

$$d = c \times \Delta t$$

$$= (3 \times 10^8 \text{ m/s}) \times (6.04 \times 10^5 \text{ s})$$

$$= 1.81 \times 10^{14} \text{ m}$$

$$= 18.1 \times 10^{13} \text{ m}$$

The Solar System is roughly 10^{13} meters in diameter, so the AGN emitting region in this example is about 18 times the diameter of the Solar System—an astoundingly small size given that a typical AGN can be a billion times more massive than the Sun and about 1,000 times more luminous than the entire Milky Way Galaxy.

1. Suppose that you observe an AGN with a variation that takes 10 seconds to pass. How large (in meters) is this AGN? (You may be able to do this math in your head, but always show your work anyway.)

Step 2—Measuring the Size of an Active Galaxy Flare Region

Imagine that you and your observing team have been monitoring a type of AGN called a quasar for about a decade. The data you obtained between 2008 and 2020 at visual wavelengths are shown in **Figure 28.2**. It is difficult to determine the "normal" level of brightness for this quasar, as there is always some level of variability. One flare is clearly visible during 2013, with another flare starting just before the observations ended.

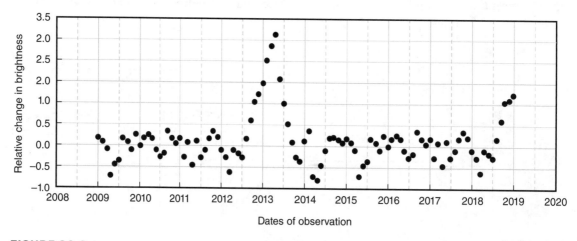

FIGURE 28.2

2. Determine the variation time, Δt, for the 2013 flare by finding the approximate start and end time (to the nearest 1/2 year) for the flare.

Start date: _____ End date: _____ Time elapsed: _____

3. Convert the variation time to seconds and round off your answer. (There are about 3.1×10^7 seconds in a year.)

$\Delta t =$ _____ seconds

4. Calculate the size of the emitting region, $d = c \times \Delta t$. (Round your answer to two decimal places.)

Size of emitting region: _____ meters

5. The Solar System is $\sim 10^{13}$ meters in diameter. What is the size of the emitting region in units of Solar Systems? Show your work.

_____ Solar Systems

Step 3—Measuring the Energy Emitted per Second by an Active Galaxy Flare

The AGN emits light in all directions, so that we receive only a tiny fraction of it here on Earth. The total energy emitted by the flare is spread out over a sphere centered on the AGN, with a radius equal to the distance from Earth to the AGN. You will calculate the total energy emitted by multiplying the power per square meter received at Earth by the total area of the sphere surrounding the AGN. This requires multiple steps.

6. The distance to this quasar is 2.4×10^9 light-years. This is the radius of the sphere over which the energy is spread out. Convert the units of this radius to meters by multiplying by 9.5×10^{15} meters per light-year (m/ly):

$$(2.4 \times 10^9 \text{ ly}) \times (9.5 \times 10^{15} \text{ m/ly}) = \text{_____} \text{ meters}$$

7. The equation for the surface area of a sphere is $A = 4\pi r^2$. Calculate the surface area using the radius of this sphere that you calculated in meters. Don't forget to square the radius and then multiply by 4π.

Area = _____ m^2

8. We estimate 1.2×10^{-13} watts per square meter (W/m^2) to be the "normal" (that is, nonflare) power per square meter received at Earth. Multiply 1.2×10^{-13} W/m^2 by the area of the sphere at Earth's distance (your answer to question 7) to find the power output of the AGN in watts.

_____ W

9. The maximum power per square meter of the 2013 flare is 3.6×10^{-13} W/m^2. What is the total luminosity of the flare? (**Hint:** Compare this number with the "normal" level in question 8. How many times greater is the maximum power per square meter from the 2013 flare than the normal level?)

_____ W

10. The Sun's luminosity is about 4×10^{26} W. What is the ratio of the maximum luminosity of the 2013 flare to the Sun's luminosity?

_____

Step 4—Putting It Together

11. Consider the observations and calculations presented here about the power emitted from the AGN and its size. Does the evidence support the hypothesis of a supermassive black hole powering your newly discovered quasar? Support your answer, using the key terms of this activity.

12. The observations presented in Figure 28.2 are nearly identical to those reported for the quasar 3C 273. Even at 2.4 billion light-years away, this quasar is one of the closest and brightest in the sky. Given the conclusion you reached for your newly discovered quasar, can you infer that a supermassive black hole is powering the quasar 3C 273? Phrase your answer in terms of a hypothesis: "If the quasar 3C 273 is being powered by a supermassive black hole, then . . ."

● ACTIVITY 29
The Distance to the Center of the Galaxy

Learning Goals

In this activity, you will use observations of RR Lyrae variable stars to determine the distances to globular clusters and from that find out where they are located in the Milky Way. You will be determining the size of the Milky Way using a method similar to that done originally. After completing the activity, you will be able to

1. determine the distances to two globular clusters using RR Lyrae variables found in them.

2. estimate the distance to the center of the Milky Way based on the distribution of 130 globular clusters.

3. relate how RR Lyrae variable stars have changed our knowledge of the Milky Way and our location in it.

Key terms: globular cluster, variable star, RR Lyrae star, average absolute magnitude, average apparent magnitude, absolute magnitude, magnitude equation

Step 1—Background

The Milky Way is shaped like a flat disk surrounded by a spherical halo. The halo is composed primarily of **globular clusters**—groups of hundreds of thousands of stars bound together by the total gravity from each individual star. Maps of the positions of these globular clusters can be used to find both the size of the Milky Way and the Sun's location within it. To map the globular clusters from three-dimensional space onto a two-dimensional sheet for this activity, we use the projected distances rather than the actual distances. As **Figure 29.1** shows, the actual distances are greater than those given here, but the center of the distribution will still be in the same location.

Distances to globular clusters are found by observing **variable stars** called **RR Lyrae stars**. Graphs of our observations of the brightness versus time for an RR Lyrae star all have a very particular shape, as shown in **Figure 29.2**, making it easy to determine which stars are RR Lyrae stars.

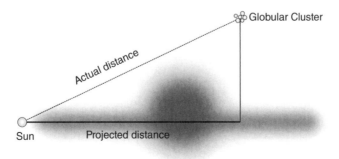

FIGURE 29.1

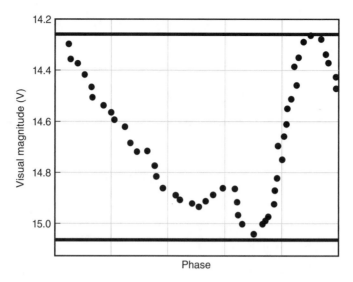

FIGURE 29.2

Light curves of six stars in the globular cluster M15 are shown in **Figure 29.3**. Light curves of eight stars in the globular cluster RU 106 are shown in **Figure 29.4**. In both figures, the *y*-axes show apparent magnitude as measured through a V (visual) filter. (Remember that the magnitude scale runs backward, so small numbers are at the top, and large numbers are at the bottom.) The observations have been "folded" over time to show a long period of time over just one or two complete cycles along the *x*-axis, what astronomers call "phase." These plots are of real observational data. Each data point represents a measurement of the brightness of that star at a certain time.

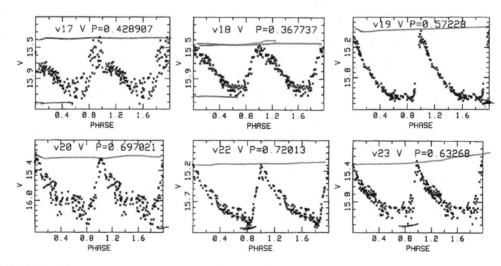

FIGURE 29.3

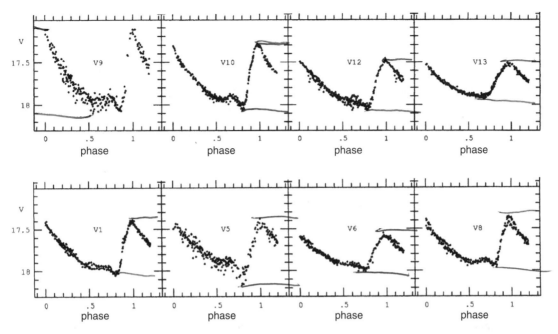

FIGURE 29.4

1. Study Figures 29.3 and 29.4, and then determine what all these light curves have in common. Carefully describe the light curve of an RR Lyrae star as though you were explaining to a friend how to determine whether a star was an RR Lyrae star.

Step 2—Finding Distances to a Clusters with RR Lyrae Variable Stars

The variable stars listed in **Tables 29.1** and **29.2** all have the same **average absolute magnitude** (same average luminosity), but because they are variable stars, their brightness changes over time. To find the **average apparent magnitude** for each star, draw a horizontal line across the top of each light curve in Figures 29.3 and 29.4 that measures the brightest magnitude. Draw another horizontal line across the bottom of each light curve that measures the dimmest magnitude. Add these magnitudes together and divide by 2 to find the average apparent magnitude. An example light curve is shown in Figure 29.2. The top horizontal line crosses the brightest magnitude: 14.25. The bottom horizontal line crosses the dimmest magnitude: 15.05. The average apparent magnitude is then (15.05 + 14.25)/2, or 14.65. Remember that the magnitude scale is backward, so a brighter measurement has a smaller number.

2. Complete Tables 29.1 and 29.2 by finding the average apparent magnitude of each of the remaining two stars in each figure.

3. Take the average of the values in the fourth column to determine the average magnitude for all the RR Lyrae stars (*m*) in each cluster. Write this at the bottom of column four in each table.

Learning Astronomy by Doing Astronomy Second Edition

TABLE 29.1

RR Lyrae stars in M15

VARIABLE STAR ID	BRIGHTEST MAGNITUDE	DIMMEST MAGNITUDE	AVERAGE MAGNITUDE
v17	15.35	16.2	15.78
v18	15.55	16.13	15.84
v19			
v20	15.2	16.5	15.85
v22			
v23	15.2	16.1	15.65
Average magnitude for M15			$m =$ 15.70

TABLE 29.2

RR Lyrae stars in RU 106

VARIABLE STAR ID	BRIGHTEST MAGNITUDE	DIMMEST MAGNITUDE	AVERAGE MAGNITUDE
v9	17.1	18.1	17.6
v10	17.25	18.1	17.68
v12	17.4	18.05	17.73
v13			
v1	17.43	18.08	17.76
v5	17.35	18.2	17.78
v6			
v8	17.33	18.0	17.67
Average magnitude for RU 106			$m =$ 17.703

Now that we know the average apparent magnitude for the stars in each globular cluster, we need to have more information in order to determine the distances to these clusters. Globular clusters are much too far away (the closest one is about 2,400 parsecs or 7,800 light-years away) for any measurements of the parallaxes of their stars. What we have discovered is that RR Lyrae stars all have similar **absolute magnitudes**. This means that they all emit nearly the same amount of light; or, in other words, that they all have the same luminosity. Once we know the luminosities and measure the brightnesses, we can find the distances.

Because we have average values for both the apparent magnitudes for each set of stars and the average absolute magnitude (M) of 0.6 for RR Lyrae stars, we can use a version of the **magnitude equation** to solve for the distance in parsecs (pc):

$$d = 10^{[(m-M+5)/5]}$$

How to solve:

a. Work with the values within the parentheses first: $(m - M + 5) =$ _____

b. Divide that number by 5 to get the exponent: _____

c. Use the 10^x button on your scientific calculator to find the distances in parsecs.

4. Use the distance equation to find the distance to M15: __6886__ pc

5. Use the distance equation to find the distance to RU106: __7633__ pc

Step 3—Finding the Distance to the Center of the Milky Way

The positions of 130 of the ~150 known Milky Way globular clusters are shown as black dots in **Figure 29.5**. This is a polar graph, so it is organized such that the two coordinates used are distance from the center and angle around from "east"; the position of the traditional positive x-axis. Each circle is 1 kiloparsec (kpc = 1,000 pc) farther from the center, and each radial line represents 15 degrees of galactic longitude. For example, the globular cluster designated by a star symbol is located at galactic longitude 180 degrees, about 11 kpc from the center. The dots at the edges of the graph represent clusters that are farther away than 19 kpc. The farthest cluster is at a projected distance of 36 kpc. The Sun and Earth are at the very center of this graph.

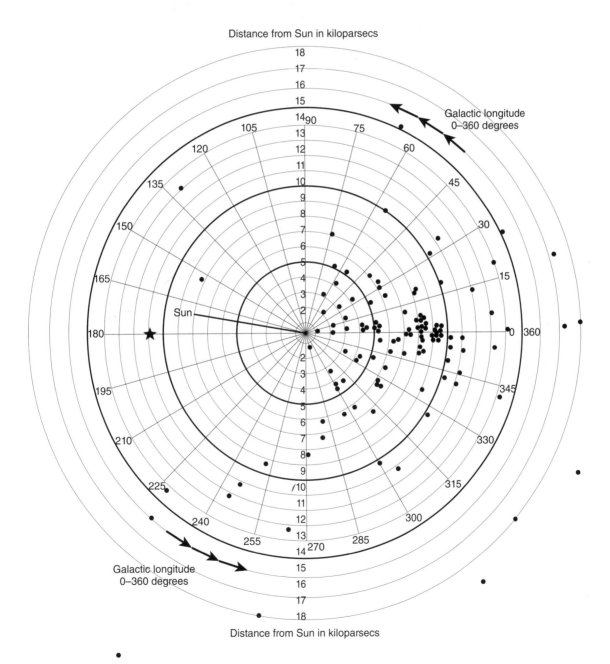

FIGURE 29.5

6. The galactic longitude of M15 is 65°; the galactic longitude of RU 106 is 301°. Based on these coordinates and the distances you found in Step 2, mark the location of M15 and RU 106 on Figure 29.5.

7. Find the center of the distribution of globular clusters. One way to do this is to use a small circular object about the size of a quarter or water bottle cap and try to include as many globular clusters as possible within the circumference of the object you use. Make an "X" on the graph at the center of the distribution.

8. Estimate the distance from the Sun to the center of the distribution of clusters.

Distance = _____ kpc

9. Estimate the radius of the Milky Way based on the full extent of the distribution of clusters. (**Hint:** Look at the farthest distances for the clusters on the graph. Your answer will be approximate.)

Approximate radius of Milky Way = _____ kpc

10. Determine the direction to the center of the distribution of clusters. This is the direction to the center of the galaxy.

Longitude to center of the distribution = _____ degrees

Step 4—Putting It Together

11. The distance to the center of the Milky Way is given as 8,122 ± 31 parsecs (8.122 ± 0.031 kpc). Compare the distance you determined from the distribution of the globular clusters to the published value (you do not need to bring in the uncertainty to do this). What percent difference did you find? Recall that the equation for percent difference is:

$$\frac{your\ value - true\ value}{true\ value} \times 100\% = \underline{\ \ \ \ }\%$$

Do you think the uncertainty in your distance value is about the same percentage? Comment.

12. Based on what you have learned through this activity, and using the key terms, explain what it is about RR Lyrae stars that gives us confidence we can use them to find distances to globular clusters. Relate the contributions RR Lyrae variable stars have made to our knowledge of the galaxy and explain how we know the Sun is not at the center of the Milky Way Galaxy.

● ACTIVITY 30
Finding the Age of the Universe

Learning Goals

In this activity, you will analyze the distances to galaxies and the velocities at which they are receding from us. By the end of this activity, you will be able to

1. explain the standard-ruler method of determining distances to galaxies and compare its measurements to those determined by redshifts.

2. use Hubble's law to estimate an age for the universe and compare it with the age of the Sun and the Milky Way.

3. summarize how Hubble's law can be used to find distances to galaxies extremely far away.

4. explain why Hubble's law implies that the universe is expanding.

Key terms: standard ruler, angular size, absorption line, emission line, redshift, blueshift, Hubble's law, Hubble constant

Step 1—Background

There are three possibilities for the how the universe changes with time: it might be expanding, it might be contracting, or it might be static—neither expanding nor contracting but staying the same. A study of the distances and velocities of nearby galaxies can be used to find out which of these is correct. If all the galaxies move at the same speed, the universe is static. If the galaxies all appear to move away from us, and the more distant ones move faster, the universe is expanding. If the galaxies all appear to move toward us, and the more distant ones move faster, the universe is contracting. A graph of the velocities, v, versus the distance, D, as shown in **Figure 30.1**, gives enough information to determine which case is correct.

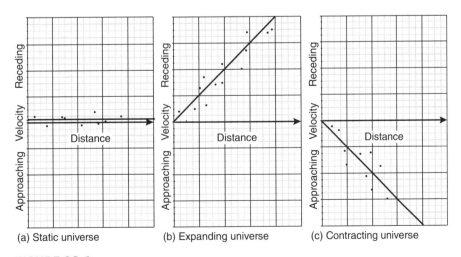

(a) Static universe   (b) Expanding universe   (c) Contracting universe

FIGURE 30.1

The distances to these galaxies can be found in a number of different ways. Here, we constrain the sample of galaxies to be all the same type—spirals—and assume they are all the same size. This is called the "**standard ruler**" assumption: for images taken with the same telescope and detector, a galaxy that looks smaller than another galaxy is, in fact, farther away. This is one of the methods that your brain uses to figure out how far away things are—if the cow looks very small, it must be very far away. This apparent size is called the **angular size**. If you were to measure an angle from one side of the object to the other, an object's angular size will be smaller if it is farther away.

Two options were available for the type of galaxy to choose: spiral and elliptical. Spiral galaxies are round like a pizza—they have a disk, with lots of dust, gas, and young stars. Elliptical galaxies are more round like a ball, and have little dust or gas, and mostly old stars. Spiral galaxies were chosen determining distances for two reasons: (a) elliptical galaxies range in size from more than a million light years across to one-tenth the diameter of the Milky Way and (b) there are many spiral galaxies close enough to the Milky Way that astronomers can use an independent method to determine their distances. This helps verify that the standard ruler assumption can be used.

Accurate velocities (whether away from or toward us) are easier to measure than distances. We find the velocities of these galaxies by measuring the Doppler shift of absorption or emission lines in their spectra. (Recall that **absorption lines** and **emission lines** are caused by electrons jumping between energy levels in atoms.) If the lines are shifted toward longer wavelengths, the shift is called a **redshift**, and the galaxy is moving away from us with a velocity greater than zero. If the lines are shifted toward shorter wavelengths, the shift is called a **blueshift**, and the galaxy is moving toward us, and the velocity is negative. In either case, the amount of the wavelength shift is proportional to the velocity of the galaxy: faster galaxies have larger shifts.

In 1929, Edwin Hubble carried out this experiment by measuring the velocities of two dozen galaxies and found that a line fit to the data increased with distance, like the line shown in Figure 30.1b. This indicated that, in general, galaxies move away from the Milky Way, and that more distant galaxies were moving away faster than closer galaxies. His conclusion that the universe is expanding is known as **Hubble's law**, and the slope of the line is known as the **Hubble constant**.

Hubble's law: The velocity of a galaxy is proportional to its distance.

$$v = H_0 d$$

If we assume that the universe has always been expanding at the same rate, then it is surprisingly straightforward to find the age of the universe from the Hubble constant: divide one by the Hubble constant, and adjust the units. A steep slope means that the universe has been expanding quickly, and the universe is young. A shallow, or flatter, slope means that the universe has been expanding slowly, and the universe is old.

1. When using the standard ruler assumption, why is it important that all of the galaxies are of the same or very similar type?

Step 2—Analyzing a Small, Representative Sample of Galaxies

We start with a closer look at a representative sample of four galaxies. Look for similarities among the images of galaxies and in the features of their spectra in **Table 30.1**. (NGC stands for New General Catalog, a catalog of clusters, nebulae, and galaxies.)

⬤ TABLE 30.1

Identification, images, and spectra of four representative galaxies. The vertical line above each *x*-axis indicates the wavelength of the hydrogen Hα line.

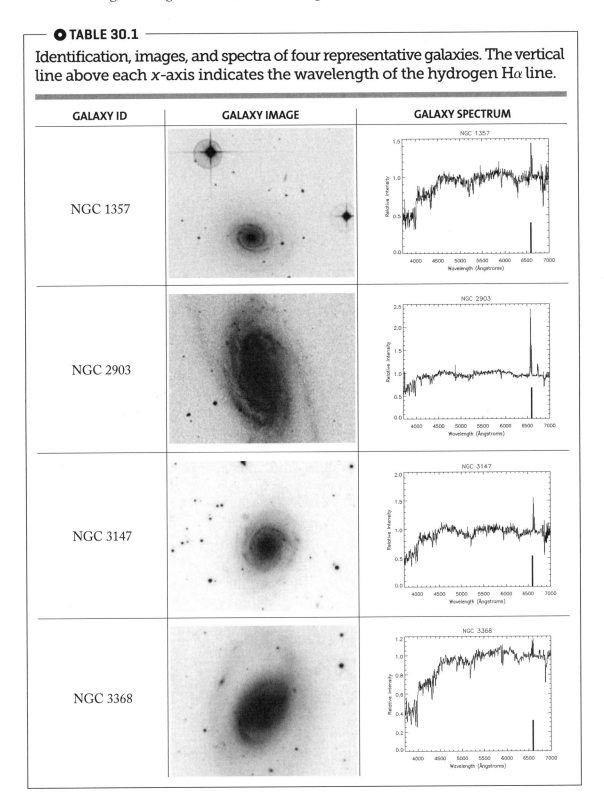

GALAXY ID	GALAXY IMAGE	GALAXY SPECTRUM
NGC 1357		
NGC 2903		
NGC 3147		
NGC 3368		

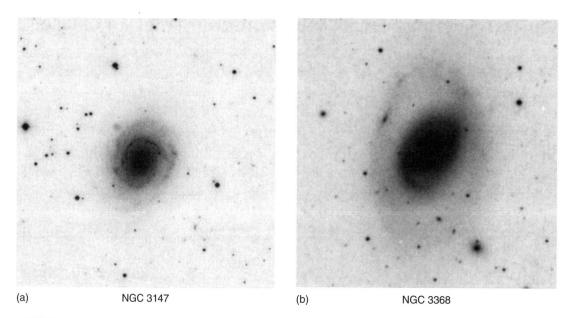

(a)   NGC 3147   (b)   NGC 3368

FIGURE 30.2

Enlarged images of two galaxies from Table 30.1 are shown in **Figure 30.2**. Each image shows a patch of sky the same size, 0.12 degrees (7.2 arc minutes) on each side. The galaxies NGC 3147 and NGC 3368 look very similar, but definitely have different angular sizes.

2. There are two probable reasons why they are not the same angular size. Either they are at the same distance and have different actual sizes, or they are the same actual size but are at different distances. Can you determine which of those two reasons is the right one? Which one of these two reasons are we assuming is true under the standard ruler method?

Hubble's law relates the velocity and the distance of a galaxy. If galaxy A is moving twice as fast as galaxy B, then galaxy A is twice as far away as galaxy B; if galaxy A is moving five times as fast as galaxy B, then galaxy A is five times farther away than galaxy B.

3. Compare the distances to these two galaxies as determined by the standard ruler method: NGC 3147 is 22 Mpc from us, while NGC 3368 is 15 Mpc from us. From this method, approximately how many times farther away is NGC 3147 than NGC 3368?

All of these galaxies are spirals, but we can add additional criteria to increase confidence that these galaxies are similar enough to be used as standard rulers. Notice that there are strong emission lines, those that show big spikes in a spectrum, of the hydrogen-alpha (Hα) line at 656.3 nm (about 6,600 Ångstroms). The wavelength of the Hα line is marked with a tick mark on the x-axis of each spectrum.

4. Strong hydrogen emission lines indicate that stars are currently forming. Hot, young, massive O and B stars emit high-energy ultraviolet photons that excite the hydrogen gas in the star-forming region. How do the hydrogen lines in the spectra of the galaxies add to our

confidence that they are spiral galaxies and not elliptical galaxies? Consider the characteristics of both spirals and elliptical galaxies from the background section in your answer.

So that we better understand how wavelength shifts of lines are interpreted, we have shown a portion of the spectra for the two galaxies of Figure 30.2 in **Figure 30.3**. These spectra highlight the regions around the absorption lines (historically called calcium H and K lines) for the singly ionized calcium atoms in the stars in these galaxies. (The rest of the "squiggles" in the spectra are due to other elements in the stars of these galaxies.) We use the Doppler formula to determine the wavelength shifts:

$$\text{wavelength shift} = \frac{\lambda_{\text{measured}} - \lambda_{\text{rest}}}{\lambda_{\text{rest}}} = \frac{v}{c},$$

where $\lambda_{\text{measured}}$ is the wavelength we measure from a spectrum (identified directly below the calcium absorption lines in Figure 30.3), λ_{rest} is the wavelength we measure in our laboratories (identified by the lines marked Ca K and Ca H on the x-axis), v is the velocity, and c is the speed of light.

5. Are these galaxies moving _____ **toward** or _____ **away** from us?

6. The galaxy NGC 3147 is moving faster than NGC 3368. How can you tell?

7. The wavelength shifts of these galaxies are positive, making them redshifts. The redshift of galaxy NGC 3147 is 0.00935, and the redshift of NGC 3368 is 0.00301. From the Doppler formula we know that wavelength shifts are proportional to the velocities of objects.

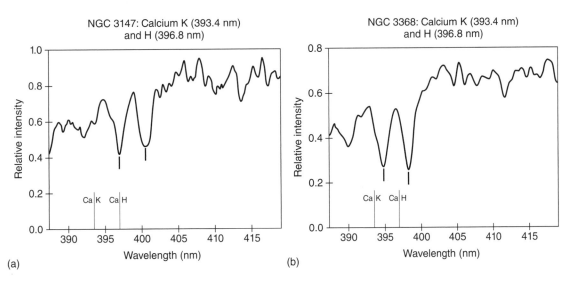

(a)   **(b)**

FIGURE 30.3

Approximately how many times faster is NGC 3147 than NGC 3368? (Take a ratio of the redshifts and state the results.)

8. How many times farther away is NGC 3147 than NGC 3368?_____ times

9. Is there an inconsistency between the relative distances obtained from the redshifts of NGC 3147 and NGC 3368 and the relative distances calculated by the standard ruler method? (See question 3.) Explain your answer.

10. Think about the underlying assumptions and methods used for the standard ruler and the direct measurements of wavelengths of lines in spectra. Which method is likely to give more accurate results? Explain.

Step 3—Graphing the Data and Finding the Hubble Constant

A graph of the measured recessional velocities versus the measured distances to 14 galaxies is shown in **Figure 30.4**. The line b is the approximate best fit to the data; lines a and c were obtained by ignoring a couple of outliers in the measurements. The equation of the line is in the form $y = mx + b$, where y equals the velocity, x is the distance, and the slope, m, will be the value of the Hubble constant. The y-intercept, b, in the equation equals zero. This equation then leads to Hubble's law: $v = H_0 d$, where v is the recessional velocity of the galaxy at the distance d.

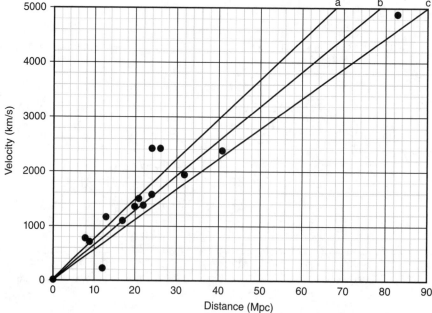

Hubble's law using 18 spiral galaxies and the standard ruler assumption

FIGURE 30.4

Learning Astronomy by Doing Astronomy Second Edition

11. Using the data in **Table 30.2**, add the four sample galaxies to the graph in Figure 30.4. Label each of the four points with the Galaxy ID.

⚫ TABLE 30.2

Distances and recessional velocities for the four sample galaxies

GALAXY ID	DISTANCE (MPC)	VELOCITY (KM/S)	GALAXY ID	DISTANCE (MPC)	VELOCITY (KM/S)
NGC 1357	25	1994	NGC 2903	10	556
NGC 3147	22	2804	NGC 3368	15	903

12. Why must the *y*-intercept in this graph be equal to zero? (**Hint**: Look at what the axes are measuring. What location in the universe is at zero distance from us? How fast does that location move away from us?)

Slopes based on the data:

Steepest reasonable a = 5000 km/s / 68 Mpc = 74 km/s/Mpc

Best-fit line b = 5000 km/s / 78 Mpc = 64 km/s/Mpc
(Use this value for the "best fit" age calculations.)

Flattest reasonable c = 5000 km/s / 90 Mpc = 56 km/s/Mpc

13. Which line will give the youngest age for the universe? _____

14. Which line will give the oldest age for the universe? _____

Step 4—Finding the Age of the Universe

We can now calculate the age of the universe by taking the inverse of the Hubble constant and adjusting the units. Study the following example to see how it is done, and then complete the calculations given in steps 15 through 17 for the best-fit slope, b.

Example: Here is an example of the math for steps 15 through 17. Suppose you find a Hubble constant of 75 km/s/Mpc, then:

For step 15, take the inverse:

$$\frac{1}{75 \text{ km/s/Mpc}} = 0.0133 \frac{\text{s} \cdot \text{Mpc}}{\text{km}}$$

For step 16, convert the distance units:

$$\left(1.33 \times 10^{-2} \text{ s} \cdot \frac{\text{Mpc}}{\text{km}}\right) \times \left(3.09 \times 10^{19} \frac{\text{km}}{\text{Mpc}}\right) = 4.11 \times 10^{17} \text{ s}$$

For step 17, convert from seconds to years:

$$(4.11 \times 10^{17} \text{ s}) \times \left(\frac{1 \text{ yr}}{3.16 \times 10^7 \text{ s}}\right) = 1.3 \times 10^{10} \text{ yr}$$

$$= 13 \text{ billion years}$$

Notice that we can move the decimal point such that 1.3×10^{10} years is the same as 13×10^9 years. In everyday language, we use the word "billion" to represent 10^9, so this age is 13 billion years. A universe with a Hubble constant of 75 km/s/Mpc is 13 billion years old.

15. Now, carry out the analysis for the value of H_o from these data. First, find the inverse of H_o by dividing 1 by the value of the slope of the best-fit line b:

Inverse: _____ s · Mpc/km

16. Second, multiply the inverse by $(3.09 \times 10^{19}$ km /Mpc) to cancel the distance units.

Age: _____ s

17. Since you now have the age of the universe in seconds, divide this number by the number of seconds in a year: 3.16×10^7 s/yr). Round off your answer.

Age: _____ years

18. Compare this age for the universe to the age of the Sun (5 billion years) and to the age of the oldest stars in the Milky Way (approximately 12.5 billion years). Be quantitative in your answer; for example, approximately how many times older or younger is the universe than the Sun? Use these comparisons to comment on whether the age that you have calculated for the universe seems reasonable.

Step 5—Putting It Together

19. You have a value for the Hubble constant from these data and from that have determined the "best fit" age of the universe. We can get a lot more information from Hubble's law. Explain how you would now go about finding the distances to galaxies that are much farther away, too far away to use the standard ruler method. Start with this form of Hubble's law: $d = {}^{v}/{H_0}$.

20. The long-standing view of the universe before Edwin Hubble's observations was that the universe was static: neither expanding nor contracting. Use at least three of the key terms for this activity to discuss how your analysis either supports or refutes this claim.

● ACTIVITY 31

Gravitational Waves and Merging Black Holes

Learning Goals

In this activity, you will use gravitational wave observations and artificial data extracted from simulations to characterize the merging of two black holes. After completing the activity, you will be able to

1. identify the best match between real data and simulated data.

2. apply conservation of energy to determine how much energy is "lost" due to both radiation and gravitational waves created from the merging system.

3. compare the results to the energy output of the Sun.

4. compare results to those published in astronomical journals.

Key terms: gravitational wave, LIGO, strain, maximum amplitude, period

Step 1—Background

Gravitational waves are ripples in spacetime caused by moving objects with mass. In 1915, Einstein predicted that gravitational waves should exist, but when he calculated the size of their effect on objects, he concluded that they would never be observed. It took experimenters almost 100 years, but in September 2014, gravitational waves were detected for the first time by the **LIGO** (Laser Interferometer Gravitational-Wave Observatory). These waves were produced by two black holes spiraling together and merging into one object.

Gravitational waves pass through the LIGO detector, alternately stretching and shrinking two long tunnels that are perpendicular to each other. Pulses of light sent down one tunnel reflect from a mirror at the end, and they return slightly sooner or later than pulses sent down the other tunnel. This difference is what scientists use to determine that a gravitational wave has passed by. Scientists express this difference as the **strain**, which measures how much every meter of the tunnel expands or contracts as the gravitational wave passes by.

As two black holes spiral together, the energy of the gravitational waves they emit increases, so they cause a larger strain to be observed. This strain varies from positive to negative and back again as the wave passes by, stretching and compressing the tunnel. The maximum amount of stretch from the unstretched position is called the **maximum amplitude**. The time it takes for the strain to go from maximum to minimum to maximum again is the **period**, even though that time is not always the same.

Several aspects of the system affect the strain, such as the angle of the binary system, whether the two black holes have similar or different masses, and whether the black holes are spinning. For this activity, we will assume that the two black holes have the same mass and are not spinning. At the end of the activity, you can compare your results with those from a more careful analysis and see how much these assumptions matter. The distance to the pair of black holes affects the amplitude of the changes; more distant systems will have smaller amplitudes. The total mass involved affects how quickly the merger happens; more massive black holes are drawn together more forcefully, and so they merge more quickly. You will investigate the effect of the distance and total mass in this activity.

1. What two assumptions will you be making in this analysis?

Step 2—Studying Models

Model predictions of how the strain will vary over time are shown in **Figure 31.1**. Each graph shows a different pair of values for distance to the colliding black holes and total mass. This is an example of a system in which multiple variables are important. To tease apart the effects of each variable, it's helpful to study them one at a time. In the left column, the total mass is the same for each graph, but the distance to the system varies. Conversely, in the right column, the distance is same for each graph, but the total mass varies. Study these eight graphs to answer the following questions.

2. How does the total mass affect the amplitude of the strain?

3. How does the distance to the system affect the amplitude of the strain?

4. How does the total mass affect the period of the variations?

5. How does the distance to the system affect the frequency of the variations?

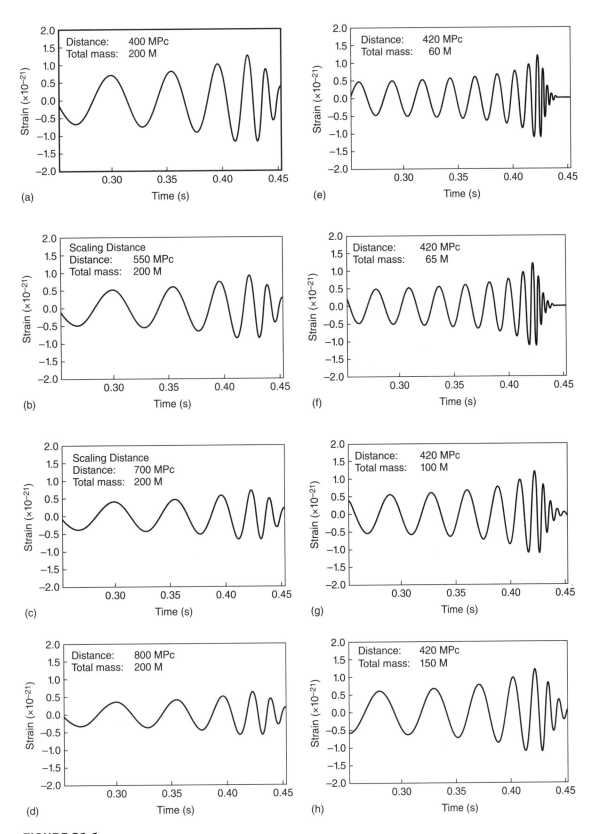

FIGURE 31.1

The first black hole merger ever detected (also the first detection of gravitational waves) took place in 2014. This event is known as GW150914; "GW" is for "gravitational wave detection," followed by the date of detection, September 15, 2014. The data from this detection are shown in **Figure 31.2**. Compare these data to the panels in Figure 31.1.

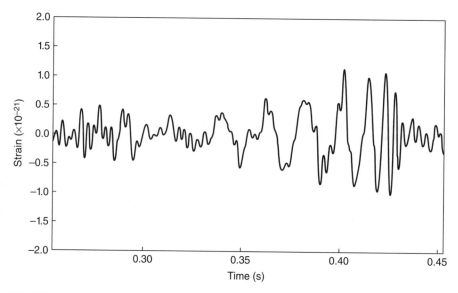

FIGURE 31.2

6. Which model (a–d) is the best fit for the amplitude of these data?

7. Which model (e–h) is the best fit for the frequency of these data?

8. Which model (a–h) is the best overall fit for these data?

9. From your analysis, what is the distance to the source of GW150914?

10. From your analysis, what was the total mass of the black holes before they combined?

It is much easier to see whether data and models match if we plot them on top of one another. This has been done in **Figure 31.3**. In this case, an additional step was taken in the model that accounts for the sensitivity of the instrument.

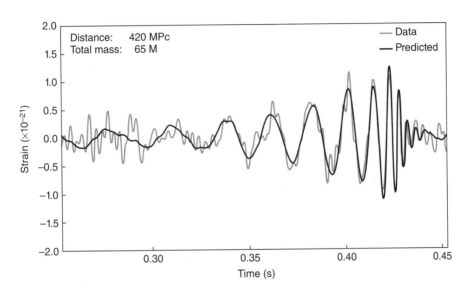

FIGURE 31.3

11. Comment on the agreement between the predicted model and the data.

Step 3—Mass and Energy

Above, you have determined that the total initial mass involved in the binary black hole merger is 65 M_{Sun}. Astronomers determined the final black hole mass to be 62.3 M_{Sun}. Some of the mass has been converted to radiation and to gravitational waves.

12. Calculate how much mass was lost when the black holes merged by subtracting the final mass from the initial mass.

Mass lost: _____ M_{Sun}

13. Use the mass-energy relation, $E = mc^2$, to calculate how much energy was released (in joules) when these two black holes merged. (Recall that $M_{Sun} = 1.99 \times 10^{30}$ kg, and $c = 3 \times 10^8$ m/s).

Setup: $E =$ (mass lost) × (mass of Sun) × (speed of light)^2

$E =$ _____ J

14. This really big number is difficult to understand. To help with that, you could compare it to the amount of energy released by the Sun over its lifetime. The Sun's luminosity is 3.38×10^{26} W. This means it emits 3.38×10^{26} joules every second. How much energy does the Sun emit in a year?

Setup: $E =$ (Sun's luminosity per s) × (3.16×10^7 s/yr)

$E =$ _____ J/yr

15. Assuming the Sun's luminosity is constant, about how much energy does the Sun emit over its 10-billion-year lifetime?

$$\text{Setup: } E = (\text{Sun's luminosity per year}) \times (10^{10} \text{ years})$$

$$E = \underline{\hspace{4cm}} \text{ J}$$

16. Compare the amount of energy released by the Sun over its lifetime to the amount of energy released in the black hole merger. Which one is larger, and how many times larger is it?

Step 4—Putting It Together

17. Using the key terms of this activity, describe how astronomers use theoretical models of gravitational waves along with observations of gravitational waves to determine the properties of the colliding black holes that emit them.

● ACTIVITY 32
The Hubble Deep Field North

Learning Goals

In this activity, you will use the Hubble Deep Field North to explore the distribution and evolution of galaxies in the universe. As you work through the activity, you will

1. analyze the galaxies in, and the structure of, the Hubble Deep Field North.

2. identify and label the Hubble types of spiral, elliptical, and irregular galaxies.

3. relate the colors of galaxies to the possible dominant stellar populations.

4. summarize the importance of the Hubble Deep Field North to our understanding of the evolution of galaxies.

5. explain how the redshifts of the galaxies give us the three-dimensional structure of the universe.

Key terms: observable universe, spiral galaxy, elliptical galaxy, irregular galaxy, absorption line, emission line, redshift, Hubble's law, Hubble constant, histogram

Step 1—Background

In this activity, you will examine the Hubble Deep Field North (HDFN), an image of very distant space as it was far back in time. This image contains a quite small portion of the sky. Imagine that you held a dime at arm's length. The eye of the president on the dime is about the same angular size as this image taken by the Hubble Space Telescope of the sky.

After this image was obtained, the 10-meter, Earth-based Keck telescope was used to observe faint blue objects in the image, objects that may represent galaxies caught as they are just forming. In all, the large number of galaxies in this very small area of the sky implies that there are about 40 billion galaxies in the **observable universe**, that part of the universe where light has had time to reach us.

Galaxies are generally classified as spiral, elliptical, or irregular types. These galaxy types are distinguishable by their shapes: **spiral galaxies** have a disk structure, often with visible spiral arms; **elliptical galaxies** look more like footballs; and **irregular galaxies** are (as the name implies) irregular in shape, being neither spiral nor elliptical.

The Doppler shift of **absorption** and **emission lines** from atoms in the galaxy can be used to determine whether a galaxy is moving toward or away from us, and by how much. The galaxies that are found in the HDFN are all far enough away from us that their lines are shifted to longer wavelengths, meaning **redshifted**. The redshift for each galaxy is measured by the observed change in the wavelengths of lines in its spectrum and is proportional to its velocity.

We use the regular Doppler shift equation for finding the velocities of these galaxies:

$$\frac{\triangle\lambda}{\lambda_{rest}} = \frac{v}{c},$$

where $\triangle\lambda$ is the measured change in the wavelength of the spectrum line, λ_{rest} is the rest or laboratory wavelength of that line, and $\frac{v}{c}$ is the velocity, v, compared to the speed of light, c. Cosmologists have introduced a new symbol for $\frac{v}{c}$: they have assigned this ratio the letter z, making $z = \frac{v}{c}$. When cosmologists list values of the redshifts for distant galaxies, instead of stating that redshift = 0.015, they might list $z = 0.015$, $z = 0.024$, $z = 0.0049$, or even $z = 1.52$, 2.8, or 4.1. Including z in our equation for the Doppler shift, and rearranging it a bit, gives us $v = c \times z$.

Edwin Hubble showed that this velocity, in turn, is proportional to distance. This proportional relationship, $v = H_0 \times d$, is known as **Hubble's law**. The constant of proportionality (H_0) is called the **Hubble constant**. Galaxies with a high redshift are located very far away; galaxies with a small redshift are (relatively) nearby.

1. Suppose that you observe two galaxies with the same redshift. What can you conclude about the distances to these two galaxies?

Step 2—Galaxy Types

Turn to **Figure 32.1** in the appendix, which shows the HDFN, and examine it carefully. Written next to many of the galaxies is the redshift, z, for that galaxy. There are a number of clearly identifiable galaxy types included in the HDFN. (You may want to refer to the source image as you work: http://ned.ipac.caltech.edu/level5/Deep_Fields/mirror/hdfn/index.html.) See if you can pick out some of the faint blue galaxies mentioned in Step 1 in the image.

2. Make your own map of the Hubble Deep Field North by locating 10–15 galaxies. Sketch them, placing them on the grid of **Figure 32.2** according to their locations in the image. It helps to use the stars shown on the grid as references, so find them first. (Look for diffraction spikes.)

3. Label each galaxy sketched with the type of that galaxy, and create a key on the grid for identification. For example, the map key gives a sketch of a star that represents the locations of stars in the image, stars that are part of the Milky Way.

4. The galaxies have noticeably different colors. Elliptical galaxies are known to have old, cool stars that dominate their light. Does the color of the elliptical galaxies you found in the image support their having these kind of stars? How?

Learning Astronomy by Doing Astronomy Second Edition

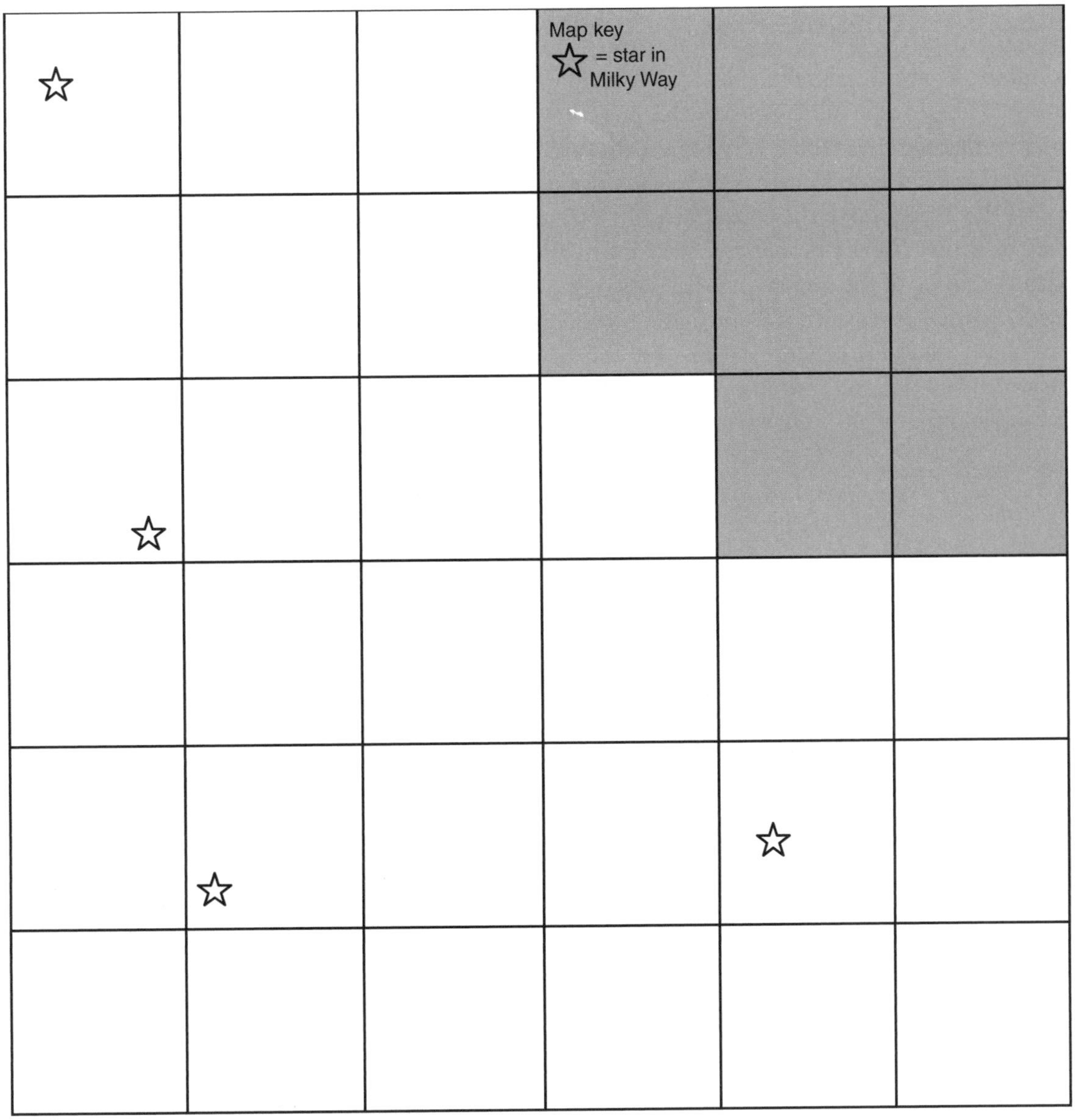

Map key
☆ = star in
Milky Way

FIGURE 32.2

Step 3—Redshifts

Return to Figure 32.1 in the appendix. You will now examine the redshifts of the galaxies in the image. The corresponding galaxy is the galaxy located closest to the redshift number.

Start by assuming the HDFN represents the actual distribution of galaxies in this portion of the sky. To determine if there is three-dimensional structure in this direction in space, you need to use the histogram of the cosmic redshifts, **Figure 32.3**. A **histogram** is a kind of bar graph. On the horizontal axis is a property of the objects, and on the vertical axis is the number of objects with that value of the property. For this histogram, each bar indicates the number of galaxies that have a particular redshift. This histogram contains data for almost all of the galaxies in this field, even those that look like little specks. For this histogram, the redshifts have been gathered in increments of 0.2, and range from $z = 0.0$ to $z = 4.2$. The curved line represents an average and emphasizes that there seem to be particular redshifts that have more galaxies than other redshifts.

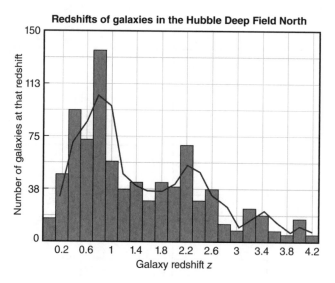

FIGURE 32.3

5. There are three redshifts for which galaxies seem to be grouped. List those redshifts.

6. At what redshifts do there seem to be very few galaxies?

7. Before developing a hypothesis, we need a question. Our question is "Why are there so few galaxies at those redshifts?" Write your own hypothesis as to why there are very few galaxies at one of the redshifts you found in question 6 by completing the following sentence: "There are very few galaxies at a redshift of _____ because ..."

8. Suppose that you had a bigger, more powerful telescope than the Hubble Space Telescope, one you could use to look at more areas of the sky and even deeper into space. How would you test the hypothesis you formed in question 7?

Step 4—The Three-Dimensional Distribution

Galaxies that appear close together on the image might actually be at very different distances from Earth. However, by adding in the redshift information, we can figure out which ones are at very different distances as well as find groups of galaxies that are physically close together in space. The second groups will be both close together in the image *and* have similar redshifts.

9. Look for a group of galaxies that are "clustered together" on the image, and sketch four or more galaxies belonging to that group of galaxies on the grid of Figure 32.2. Circle your cluster on Figure 32.2.

10. By how much do the redshifts vary among these galaxies?

11. Do you think these galaxies actually form a cluster? Why or why not?

12. There are a number of galaxies in this image that have identical redshifts; for example, about a dozen have $z = 0.56$ and a half dozen have $z = 0.48$. Suppose you noted two clusters where all the galaxies in one group had a redshift $z = 0.20$ and all the galaxies in the other group had a redshift $z = 0.40$. Use the relationship between redshift and velocity, $v = c \times z$, to determine the difference in velocities between these two clusters of galaxies. Use 300,000 km/s as the speed of light.

13. Use these velocities with a Hubble constant of $H_0 = 70$ km/s/Mpc and Hubble's law ($v = H_0 \times d$) to calculate the distances to the two clusters. Which galaxy cluster is the farther one?

Step 5—Putting It Together

Astronomers examine the images of the galaxies in the Hubble Deep Field North because there are galaxies that show how the universe was at a relatively early age. Astronomers can compare these galaxies, which are newly formed, to nearer galaxies, which are much older, to understand the changes in galaxies over time. For example, when we examine galaxies that have $z = 4.0$, we are seeing how they looked when they were only 2–3 billion years old. Examining galaxies at $z = 2$ gives us clues as to how galaxies looked when they were 3–4 billion years old. Those at $z = 1$ should give us a good idea of how galaxies looked at 10 billion years old.

14. As part of Step 1 and the mapping, classifying, and examining of the redshifts of these galaxies, you most likely noted that those with the largest redshifts looked different from those with the smallest ones. Describe how cosmologists can use these differences to study the evolution of galaxies.

15. Use key terms of this activity to explain why studying the Hubble Deep Field North should give us insight as to the three-dimensional structure of the universe.

Name _____ Date _____ Section_____

● ACTIVITY 33

Calculating the Mass of the Black Hole at the Center of the Galaxy

Learning Goals

In this activity, you will use observations of stars orbiting near the center of the Milky Way to confirm the existence of a supermassive black hole as the central object. After completing the activity, you will be able to

1. demonstrate proficiency in the use of scientific notation in all calculations.

2. apply Newton's form of Kepler's third law to infer the mass within the orbit of a centrally located star in the galaxy.

3. compare the results to the mass of the Sun and the size of the Solar System.

4. analyze the evidence for a supermassive black hole at the center of the Milky Way.

5. compare results to those published in astronomical journals.

Key terms: period, semimajor axis, black hole

Step 1—Background

Elliptical orbits can be described by two numbers: the **period** is the time it takes for the object to go around once; the **semimajor axis** is the average distance of the orbiting object from the mass that it orbits. This is also equal to half of the longest (major) axis of the ellipse, which is where the name "semimajor axis" comes from. If the orbiting object is much less massive than the object that it orbits, these two numbers are all that is needed to determine the mass of the object being orbited. This method, of course, is the one developed by Sir Isaac Newton. Our calculations using his equation have found the mass of Jupiter from the orbits of its moons, the mass of the Sun from the orbits of planets, and the mass of a supermassive black hole from the stars that orbit it. A **black hole**, whether it came from the explosion of a massive star or is located in the center of a galaxy, can be identified from observations of orbiting objects because it will have a very large mass crammed into a very small amount of space.

In fact, studies of the motions of stars closest to the Sgr A* source at the center of the Milky Way suggest a central mass enormously greater than that of the few hundred stars orbiting there. Stars closer than 0.1 light-years from the galactic center follow Kepler's laws, indicating that their motion is dominated by mass within their orbit. The closest stars studied are only about 0.01 light-years from the center of the galaxy—so close that their orbital periods are only a dozen years or so. The positions of those stars change noticeably over time, and astronomers can see them speed up as they whip around the supermassive object at the galaxy's center. To examine the motions of these stars, visit http://www.astro.ucla.edu/~ghezgroup/gc/videos/ghezGC_comp3 -18_H264_864_VP8.webm.

1. What two numbers are required to determine the mass of the object around which another object orbits?

Step 2—Finding the Mass and Semimajor Axis of the Central Object

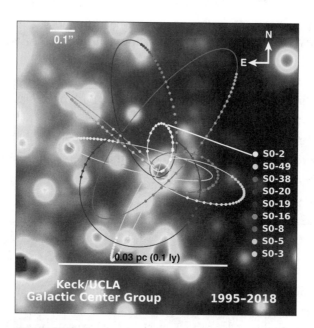

FIGURE 33.1

2. During the 18 years of observations, the star S0-2 (see **Figure 33.1**) completed a full orbit having a period of about 16 years. Convert this period to seconds (1 year = 3.16×10^7 seconds).

    Setup: 16 years $\times$ 3.16 $\times$ 10⁷ seconds per year = period, P:

    $$P = \text{_____ seconds}$$

3. Finding the semimajor axis, a:

    a. The angular size of the orbit of S0-2 is 0.165 arcsec. Convert the angular size, Θ, of the orbit in arcseconds to radians (rad), by dividing the angular size in arcsec by the number of arcsec in a radian: 206,265 arcsec/rad.

    Setup: 0.165 arcsec/2.06 $\times$ 10⁵ arcsec/rad = angular size in radians, Θ:

    $$\Theta = \text{_____ rad}$$

    b. The distance to the center of the galaxy is given as 7,940 parsecs (pc). Convert this distance to meters, by multiplying by the number of meters in a parsec: 3.09×10^{16} m/pc.

    Setup: 7,940 pc $\times$ 3.09 $\times$ 10¹⁶ m/pc = distance in meters, D:

    $$D = \text{_____ meters}$$

c. Find the major axis, d, of the orbit using the small-angle formula: $d = \Theta \times D$.

$$d = \underline{\hspace{3cm}} \text{ meters}$$

d. Find the size of the semimajor axis, A, by dividing the major axis by 2.

$$A = \underline{\hspace{3cm}} \text{ meters}$$

4. To find the mass of the object being orbited, use Newton's version of Kepler's third law, recalling the gravitational constant, $G = 6.67 \times 10^{-11} \text{ m}^3/\text{kg} \cdot \text{s}^2$.

$$\text{Setup: } M = \frac{4\pi^2}{G} \times \frac{A^3}{P^2} = 5.92 \times 10^{11} \text{ kg} \cdot \text{s}^2/\text{m}^3 \times \frac{A^3}{P^2};$$

$$A^3 = \underline{\hspace{3cm}} \text{ m}^3, P^2 = \underline{\hspace{3cm}} \text{ s}^2, M = \underline{\hspace{3cm}} \text{ kg}$$

5. The mass of the Sun is approximately 1.99×10^{30} kg. Divide the mass of the object being orbited by the mass of the Sun to express the mass of the object at the center of the galaxy in solar masses.

$$M = \underline{\hspace{3cm}} \text{ M}_{Sun}$$

6. Compare the size of the semimajor axis of the orbit of star S0-2 to the size of the Solar System, using the distance to the outer edge of the Kuiper Belt of about 55 astronomical units (AU). There are 1.496×10^{11} meters per AU:

$$(55 \text{ AU} \times 1.496 \times 10^{11} \text{ m/AU}) = 8.23 \times 10^{12} \text{ m}.$$

Divide the value of A you found in question 3d by this number to change the AU in meters to Solar System radii.

$$A = \underline{\hspace{3cm}} \text{ Solar System radii}$$

7. The size of the central object has to be smaller than the semimajor axis of the orbit of star S0-2. Comment on your results for the mass of the central object compared to the estimate of its size.

Step 3—Considering Uncertainties and Comparing Results

Our measurement for the star's semimajor axis assumed that it is not tilted from our point of view. It is entirely possible that the orbital plane of the star is tilted toward or away from us. **Figure 33.2** demonstrates this effect for the possible inclinations of the orbit of star S0-2.

How we view the orbit of star S0-2

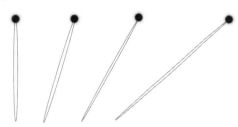

What we don't know about the orbit if we rotate our perspective vertically is the orbital tilt of the star.

FIGURE 33.2

8. If the orbit is tilted, the actual semimajor axis will be larger than what you measured. Consider the equation you used to solve for the mass of the central object. Will a tilted orbit from our perspective make the mass of the central object larger or smaller than your estimate? Explain.

9. A number of recent publications give the mass of the central object of the galaxy as approximately 4 million times the mass of the Sun. How do your results compare? Be sure to include comments about the uncertainty of the tilt of the orbit.

Step 4—Putting It Together

10. Astronomers are fairly sure that there is a supermassive black hole at the center of our galaxy. Do you agree or disagree? Support your answer with the key terms and results you discovered here.

● ACTIVITY 34
Timing from the Big Bang to Today

Learning Goals

By the end of this activity, you will be able to

1. interpret graphs of changes in temperatures and densities of the universe.

2. compare stages of the evolution of the universe to regions of the Sun.

3. sort the stages of the evolution of the universe.

4. relate the stages to a pictorial version showing the Big Bang to today.

5. summarize how the expansion of the universe led to the formation of Earth.

Key terms: cosmic microwave background radiation (CMBR), nucleosynthesis, elementary particle, annihilation, electroweak force, grand unified theory, theory of everything (TOE)

Step 1—Background

The **cosmic microwave background radiation** (**CMBR**) is the radiation emitted when the universe became cooled enough for electrons to join with nuclei to form atoms. This was when the universe became transparent for the first time, 380,000 years after the Big Bang. Analysis of the CMBR shows that the universe at that time was hot, dense, and opaque. The blackbody curve of the radiation gives us the current temperature of the universe: 2.73 K. Tiny differences in temperature of the CMBR from place to place represent the slightly denser regions where the first stars and galaxies would form. These characteristics give us the clues we need to figure out what the universe was like *before* it was 380,000 years old.

What would it be like if we could travel back to the Big Bang? Starting with the emission of the CMBR and reversing time, we pass into a universe containing mostly hydrogen and helium nuclei and freely roaming electrons. There are very few atoms. As soon as an electron joins with a nucleus to form an atom, an energetic photon ionizes it again. The universe is denser and hotter at earlier times, until there are just nuclei and electrons. As we travel back in time, it takes us nearly the entire 380,000 years to get to the next stage.

Big Bang **nucleosynthesis**, the fusion of hydrogen nuclei into helium, was ending when the universe was 15–20 minutes old. As nucleosynthesis comes to an end, the universe is made up of 75 percent hydrogen and 25 percent helium, with a tiny bit of deuterium. The temperature is 10^9 K (a billion kelvins) and the density is 10^5 kg/m^3 (100 times that of water). Going back in time reverses the fusion of hydrogen into helium, and we find just protons and electrons by the time we reach 3 minutes after the Big Bang. Continuing on, we cross through a temperature of 3 billion K and a density 10,000 times that of water.

We are now at the time where there are just the **elementary particles** with bizarre names like quarks, muons, pions, and bosons. The **annihilation** of matter and antimatter is occurring, creating energetic photons that then recombine to form matter and antimatter. Before this era, the electromagnetic force and weak force were combined into the **electroweak force**. Earlier yet, the strong and electroweak forces were one force under the **grand unified theory**.

Now we are at the time when the temperature is 10^{28} K and the density is 10^{80} times that of water. We watch the universe shrink exponentially (reversing its 10^{60} growth originally), sometime between 10^{-32} and 10^{-33} seconds after the Big Bang. It is now the Planck era, 10^{-43} seconds after the Big Bang, and the temperature is 10^{32} K and the density is 10^{96} times that of water. The four fundamental forces are combined under the **theory of everything** (**TOE**). We have arrived at a time when the conditions of the universe are unknown; our theories are unable to describe it.

1. At the stage where Big Bang nucleosynthesis was occurring, the temperature and density of the universe were about _____ K and _____ kg/m³

Step 2—Comparing Stages of the Universe to Regions of the Sun

Before we consider the timeline of the universe, let's review the structure of the Sun and investigate how its interior pressure, density, and temperature change as a function of its radius, as graphed in **Figure 34.1**. Temperatures in the interior of the Sun are similar to those of the beginning of the universe as long as we start with the time of Big Bang nucleosynthesis when more massive atomic nuclei were being formed from less massive nuclei. When this fusion of nuclei occurs in stars, it is called stellar nucleosynthesis, a slightly different term for the same process.

The innermost region of the Sun is called the core. Nucleosynthesis takes place in the core of the Sun, where the temperature is at least 1.5×10^7 K and the density at least 1.5×10^5 kg/m³. Nucleosynthesis happens only in the inner 20 percent of the Sun because the pressures, temperatures, and densities are too low in regions farther out.

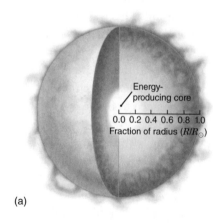

(a)

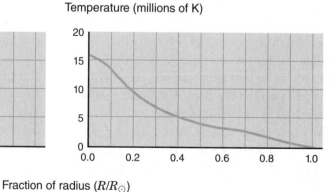

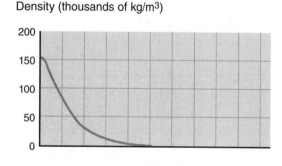

(b)

FIGURE 34.1

The next region of the Sun, the radiative zone, runs from the core to the convection zone. Temperatures in the radiative zone are not high enough for fusion but are high enough to ionize atoms into their nuclei and electrons. Energy produced in the core takes a long, meandering path through this zone. The pressure, temperature, and density finally drop low enough in the photosphere, about 4000 K, for atoms to form and persist. The photons no longer have enough energy to ionize the atoms, and the light escapes into the Solar System.

There is, of course, a big difference between the characteristics of the interior of the Sun and the universe: we can plot these quantities against the radius for the Sun, but similar processes just after the Big Bang occurred everywhere. To make a graph of how they change, we need to plot them against time. The temperature and density of the universe going forward in time are summarized in **Figure 34.2**.

How close have scientists come to replicating the conditions of the early universe? The Large Hadron Collider, located in Switzerland, has been able to reproduce the extremely hot, dense soup (called the quark-gluon plasma) that existed less than a second after the Big Bang. It does this by accelerating and colliding two beams of heavy ions (lead or gold nuclei). The temperature at one of these collisions exceeds 1.5×10^{12} K.

2. Compare Figures 34.1 and 34.2 to determine the approximate time after the Big Bang when the universe was the same

   a. temperature as the photosphere (surface) of the Sun. _____

   b. temperature as the core of the Sun. _____

   c. temperature as during a collision of heavy ions in the LHC. _____

   d. density as a neutron star (10^{17} kg/m³). _____

   e. density as a white dwarf (10^9 kg/m³). _____

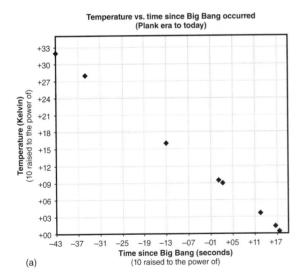

(a)

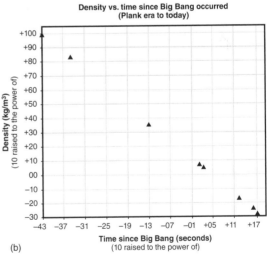

(b)

FIGURE 34.2

Step 3—Sorting the Stages of the Universe

The evolution of the universe was driven by the expansion of the universe, which caused both the temperature and the density to drop as time passed. It was impossible for each event to occur without a required previous event. For example, the fusion of hydrogen to helium during Big Bang nucleosynthesis could not occur until the protons were created, and the protons didn't exist until the annihilation of matter and antimatter stopped.

3. There is a scrambled list of the stages of the evolution of the universe given in **Table 34.1** at the end of this activity. In the space provided below Table 34.1, place these stages in the order in which they occurred, based on the information given in the background.

Step 4—Picturing the Ordered Stages of the Evolution of the Universe

4. You now have the stages in the order that you think they occurred. Write those stages into the blanks shown in **Figure 34.3**.

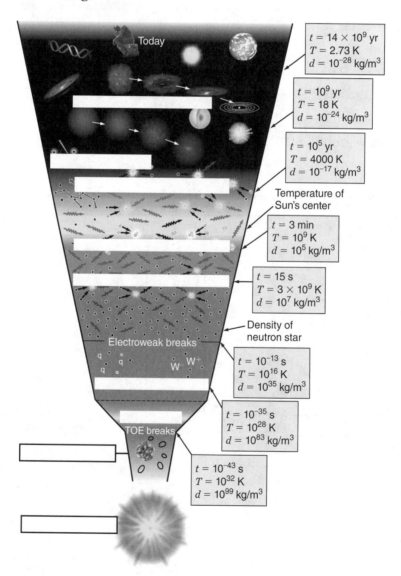

FIGURE 34.3

Step 5—Putting It Together

5. Based on the stages of the evolution of the universe that you wrote into Figure 34.3, which stage most closely matches the conditions of the

   a. core of the Sun? _____

   b. radiative zone of the Sun? _____

   c. photosphere (surface) of the Sun? _____

   d. collision chamber of the Large Hadron Collider? _____

6. "The universe started out extremely hot and tremendously dense. The expansion of the universe from that point in time was responsible for the various stages of the universe, and it is the reason we are here today." Use these two sentences as the start of a 300-word essay that summarizes what you have learned about the stages of the universe, from the Big Bang until today. Use at least four key terms in your essay.

Learning Astronomy by Doing Astronomy Second Edition

⊙ **TABLE 34.1**

Stages in the evolution of the universe

Big Bang

Big Bang nucleosynthesis

Galaxies, stars, and planets form

Inflation

Matter-antimatter annihilation

Atoms form; nuclei grab electrons

Planck era

Three fundamental forces break

Cosmic microwave background

Correct order:

Big Bang _____

_____

_____

_____

_____

_____

_____

Galaxies, stars, and planets form

● ACTIVITY 35
Finding Habitable Worlds beyond Earth

Learning Goals

In this activity, you will explore the properties of known exoplanets to determine whether they might support life. During this activity, you will also

1. describe physical and orbital characteristics of a number of exoplanets.

2. quantitatively compare characteristics of exoplanets to the planets in the Solar System.

3. apply the definition of *habitable zone* to determine habitability of exoplanets.

4. summarize what life might be like on a world in orbit around another star.

Key terms: periastron, apastron, habitable zone, exoplanet

Step 1—Background

We can determine how the temperature of a planet changes from its closest approach to its star during its orbit, **periastron**, to its farthest distance from its star during its orbit, **apastron**. When a planet is closer to its central star, it receives more light, and therefore is hotter. The planet will receive the most light, and be hottest, when it is at periastron. It will receive the least light, and be coolest, when it is at apastron. This temperature change can be expressed as a ratio (for example, the temperature at periastron is 1.2 times hotter than at apastron), or as a percentage (the temperature at periastron is 20 percent hotter than at apastron). In these calculations, we ignore any changes in the amount of light reflected by the planet (for example, because of vegetation), and since we don't know what the axial tilt of a planet is, we assume that the axes are oriented perpendicular to the orbit: "straight up and down."

Kepler's laws seem to enter into much of what we do in astronomy. We need to consider the second law: planets sweep out equal areas in equal time. To state this in a physical way: planets move the fastest when they are closest to the Sun and the slowest when they are the farthest.

A common definition of the **habitable zone** is that it is the range of distances from the central star in which liquid water might exist on the surface of a planet, if the planet has a dense enough atmosphere. Since Earth satisfies these criteria, and is habitable, the Sun must have a habitable zone.

1. Create a sentence or phrase that will help you remember the difference between periastron and apastron, and which one is which. (For example, ROY G. BIV helps you remember the order of the colors in the rainbow.) Write your sentence or phrase here:

Step 2—Stellar Properties

Information for a small sample of the **exoplanets** whose orbits are within or close to the habitable zone for their respective stars is collected in **Table 35.1**. Information for five Solar System planets is included for comparison.

⬤ TABLE 35.1

Comparison of orbital elements and stellar properties for selected exoplanets and five planets of the Solar System

PLANET NAME (DESIGNATION)	MINIMUM MASS ($M_{JUPITER}$)	RADIUS ($R_{JUPITER}$)	SEMIMAJOR AXIS (AU)	ORBITAL PERIOD (DAYS)	ORBITAL ECCENTRICITY	MASS OF STAR (M_{SUN})	TEMPERATURE OF STAR (K)
Mercury	0.0002	0.03	0.4	88	0.21	1	5778
Earth	0.003	0.09	1.0	365	0.0167	1	5778
Mars	0.0003	0.05	1.5	687	0.09	1	5778
Jupiter	1	1	5.2	4,270	0.05	1	5778
Saturn	0.3	0.8	9.6	10,759	0.06	1	5778
HD 10180 g	0.067	—	1.42	602	0.00	1.06	5911
HD 99109 b	0.50	—	1.11	439	0.09	0.94	5272
HD 28185 b	5.8	~1*	1.02	379	0.07	0.99	5656
HD 73534 b	1.07	~1	3.02	1,770	0.05	1.17	4884
HD 183263 b	3.57	~1	1.49	626	0.36	1.12	5936
55 Cnc f	0.173	—	0.77	261	0.32	0.9	5196

*It is hypothesized that Jupiter's radius is about as large as a gaseous planet can get because additional mass will tend to squeeze the planet more. For the planets in this sample with masses lower than that of Jupiter, no radius is known because they do not transit their stars.

2. Examine the range in surface temperatures of the stars with exoplanets by comparing the temperatures of the hottest and coolest stars with the temperature of the Sun (5778 K). Comment on your results.

3. Examine the range in masses of the stars with exoplanets by comparing the masses of the most massive star and the least massive star with the mass of the Sun (the mass of the Sun is 1 in these units). Comment on your results.

Step 3—Temperature and Eccentricity

Comparisons of the temperatures at periastron versus the temperatures at apastron for a sample of planets are shown in **Figure 35.1**. The vertical axis gives the ratio of the temperature at periastron to the temperature at apastron. This ratio is always bigger than one because the planet will be hotter when it receives more light at periastron. Orbits with larger eccentricities have a larger difference between periastron and apastron, so the temperature changes more between the two locations in the orbit.

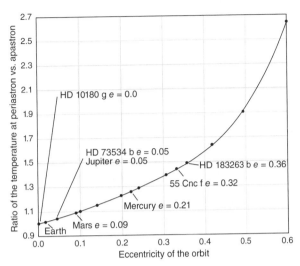

FIGURE 35.1

For example, Earth's eccentricity of 0.0167 is very small. The ratio of the periastron temperature to apastron temperature is about 1.017; Earth is 1.7 percent hotter at periastron than at apastron. (Interestingly, Earth is closest to the Sun in early January, and farthest away in early July.) As another example, the orbit of Mars has an eccentricity of 0.09. The ratio of temperatures is 1.10, so the temperature of Mars would be about 10 percent greater at perihelion than at aphelion. In contrast, the planet identified as 55 Cnc f has an orbital eccentricity of 0.32, which gives a ratio of a little over 1.4. This planet is 40 percent hotter at periastron than it is at apastron.

For the following questions, assume that only the eccentricity of Earth changes. The timing of periastron does not change, the axial tilt of the Earth does not change, nor does its average distance from the Sun.

4. On Earth, seasons are dominated by the tilt of Earth's axis, which changes the angle of sunlight throughout the year. What would the seasons be like if Earth's orbit had the same eccentricity as Mars's orbit?

5. Recall Kepler's second law: Would the Northern Hemisphere's summer be longer or shorter? Explain.

6. Would the Northern Hemisphere's summer be hotter or colder? Explain.

---

7. From Figure 35.1, find the ratio of temperatures for Mercury. How much hotter is Mercury at periastron than at apastron (express this as a percentage as in the examples given)?

8. What would humans on Earth need to do to withstand this temperature range if Earth's orbit had the same eccentricity as that of Mercury? Provide a few specific examples.

Step 4—Habitable Zones

One estimate of the Sun's habitable zone is shown in **Figure 35.2**. The light-gray shaded area depicts the conservative estimate of the habitable zone, while the dark-gray shaded areas show more optimistic estimates of the inner and outer boundaries of the habitable zone. The optimistic estimate includes an area about halfway between the orbits of Earth and Venus (inner edge) and the distance of Mars's orbit at that planet's aphelion. The conservative estimate starts just inside Earth's orbit and extends out to Mars's orbit at that planet's perihelion distance. The full details of the orbits of the selected exoplanets are listed in **Table 35.2**, along with the conservative and optimistic estimates of the habitable zones.

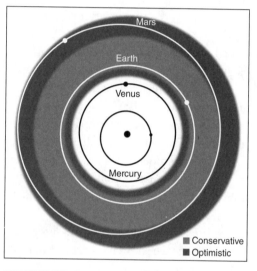

FIGURE 35.2

⦿ TABLE 35.2

Habitable zones of selected exoplanets. Identifying letters refer to the identification of the planetary system in Figure 35.3.

	PLANET DESIGNATION	DISTANCE FROM STAR (AU)	PERIOD (DAYS)	ORBITAL ECCENTRICITY	CONSERVATIVE (AU)		OPTIMISTIC (AU)	
					INNER HABITABLE ZONE	OUTER HABITABLE ZONE	INNER HABITABLE ZONE	OUTER HABITABLE ZONE
	Earth	1.00	365	0.06	0.95	1.4	0.85	1.7
(a)	HD 10180 g	1.42	602	0.00	1.13	1.94	0.87	2.11
(b)	HD 99109 b	1.11	439	0.09	0.82	1.43	0.62	1.56
(c)	HD 28185 b	1.02	379	0.07	0.95	1.65	0.73	1.80
(d)	HD 73534 b	3.02	1,770	0.05	2.14	3.78	1.61	4.13
(e)	HD 183263 b	1.49	626	0.36	1.15	1.97	0.88	2.15
(f)	55 Cnc f	0.77	261	0.32	0.77	1.35	0.59	1.48

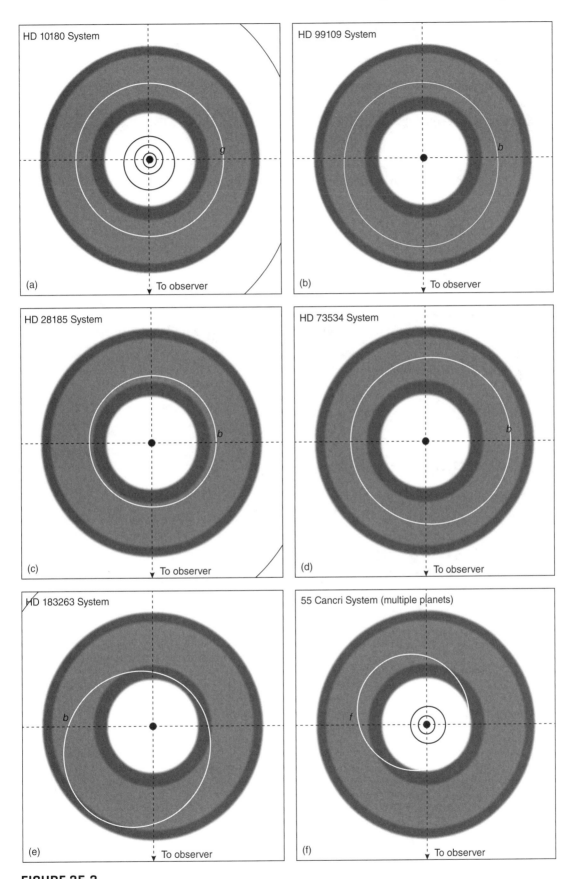

FIGURE 35.3

9. The estimated habitable zones for the six exoplanets are shown in **Figure 35.3**. Four of the exoplanets have orbits that lie completely within the conservative estimate for the habitable zone of their star. Circle the names of these planets in Table 35.2.

10. The exoplanets listed in this activity were selected because of their low orbital eccentricities and their positions within the habitable zones of their respective stars. Which one of the four exoplanets that you circled in Table 35.2 is most like Earth? In what ways is it different?

11. Earth's seasons and weather are not strongly affected by our orbital eccentricity. Which one of these four exoplanets probably experiences the largest range in temperatures based on its orbit? How would it be affected, and why?

12. Most of these exoplanets are probably gaseous, especially those close to Jupiter's mass or larger. If all of these exoplanets were to have at least one moon with a solid surface, which exoplanet's moon would you choose to visit or live on? Explain your decision.

Step 5—Putting It Together

13. Summarize, using key terms, what it would be like to live on the world that you concluded would be most habitable (question 12). You should use the information about the exoplanet's distance from its star, its orbit, its mass and radius (if known), and its probable composition and density (gaseous or terrestrial).

● ACTIVITY 36
A Cosmic Calendar

Learning Goals

In this activity, you will organize the natural history of the universe onto a calendar representing one Earth year in order to gain a feel for the relative magnitudes of the timescales involved. During the activity, you will need to

1. calculate relative timescales.
2. identify important events in natural history.
3. choose appropriate time steps at different timescales.

Key terms: scaling, Big Bang

Step 1—Background

You are familiar with **scaling** relations from maps, which transform one length scale into another in order to "de-magnify" reality so that it fits onto a piece of paper. This makes finding the relationships between different locations much simpler. We can perform a similar trick with timescales, scaling the entire age of the universe onto a cosmic calendar of only one year, so that we can more easily understand the relationships between the time intervals. In order to make a cosmic calendar, we have to reorganize our timescale so that the **Big Bang**, which occurred 13.8 billion years ago, occurs at time zero.

1. To understand the scaling of the cosmic calendar, you must first fix in your head the meaning of a "year," a "month," a "day," and so on.

   a. A lot happens in a year. Take a moment to think about the past year. List a significant event that happened to you about a year ago, and one more recent event.

   b. Now take a moment to think about the past month. List one or two events from the past month that are memorable.

   c. Think about the past day. List one or two things that happened in the past day that caught your attention.

d. For the past minute, you have been focused on this assignment. But take a moment to think about something significant that can happen between one minute and the next, and write it down, even if it didn't happen to you THIS minute.

e. Think of something significant that can happen between one second and the next, and write it down, even if it didn't happen to you THIS second.

Step 2—Scaling Dates and Times

2. Study **Table 36.1**, which lists a number of events in the history of the universe, and the time at which they occurred, in number of years ago. Column 3 lists the fraction of the year *since the beginning* at which the event happens. Fill in the empty spots in column 3 of Table 36.1 by dividing the time of the event in units of "years ago" by the total age of the universe, 13.8×10^9 years, and then subtracting that number from 1. (Notice that the value of your answer should come between the numbers immediately above and below.) For example, the Milky Way formed 11×10^9 years ago. To find the fraction of the year, we divided 11×10^9 by 13.8×10^9 to get 0.797, which is very nearly 0.8. Then we subtracted this from one: $1 - 0.8 = 0.2$.

3. Column 4 gives the day number that corresponds to the scaled time of the event. Fill in the empty spots in column 4 of Table 36.1 by multiplying the number in column 3 by 365. For the Milky Way, we multiplied 0.2 by 365 to get 73 days.

4. Column 5 shows the date corresponding to the day number from column 4. Because each month contains a different number of days, this is a little tricky. To help you with the conversion from day number to date, we've listed the day number of the first day of each month in **Table 36.2**. Fill in the empty spots in column 5 of Table 36.1. For example, for the Milky Way, we compared 73 to the day numbers in Table 36.2 to find that March 1 was day 60. So the Milky Way formed 13 ($73 - 60$) days after March 1, on March 14.

5. Column 6 shows the time (in hours, minutes, and seconds) that corresponds to the fractional day in column 4. This is only necessary on the last day of the year because so many events happened so close together in time. Time is not measured in base 10, but instead has 60 seconds in a minute, 60 minutes in an hour, 24 hours in a day, and 365 days in a year. Fill in the three blanks in column 6 by following along with the calculation for the rise of agriculture given here:

   a. Find the fractional day by subtracting 364 from the number in column 4. For example, agriculture began on day number 364.99968. This is near the end of the day on December 31. How close to the end of this last day of the year? The rise of agriculture occurred 0.99968 of the way through the day.

   b. Multiply this fraction of a day by 24 hours to find out how many hours are accounted for in this fraction. In the case of the rise of agriculture, $0.99968 \times 24 = 23.99232$ hours. Agriculture arose in the last hour of the last day.

Learning Astronomy by Doing Astronomy Second Edition

c. To find the number of minutes of the last hour of the last day, we must take the fractional hour (0.99232) and multiply it by 60, which gives 59.5392 minutes. Agriculture arose on December 31, in the 23rd hour 59th minute.

d. To find the number of seconds in the last minute of the last hour of the last day, we must take the fractional minute (0.5392) and multiply it by 60, which gives 32.352. Depending on the times around this time, we might retain the fractional second or we might not. In this case, the events on either side of the rise of agriculture are many seconds earlier or later, so we can ignore the fractional second. The rise of agriculture occurred on December 31 at 23 hours 59 minutes 32 seconds.

⦿ TABLE 36.1

Some events in the history of the universe

EVENT	ACTUAL TIME (YEARS AGO)	1 – (FRACTION OF YEAR)	DAY NO.	DATE	TIME
Big Bang	13.8×10^9	0	1	January 1	—
Milky Way formed	11×10^9	0.2	73	March 14	—
Sun formed	4.6×10^9		245	September 2	—
Oldest known Earth rocks	4.0×10^9	0.71	259	September 16	—
First (known) life	3.8×10^9	0.72	264	September 21	—
Photosynthesis	3×10^9	0.78		October 13	—
Oxygenation of atmosphere	2.4×10^9	0.83	302	October 29	—
Eukaryotes	2×10^9	0.86	312	November 8	—
Multicellular life	1×10^9	0.93	339		—
Simple animals	0.67×10^9	0.95	347	December 13	—
Fish	0.5×10^9	0.96	352	December 18	—
Amphibians	0.36×10^9	0.97	355	December 20	—
Mammals	0.2×10^9	0.986	360	December 25	—
Birds	0.15×10^9	0.989	361	December 27	—
Flowers	0.1×10^9		362	December 28	—
Primates	65×10^6	0.995	363	December 29	—
Apes	15×10^6	0.9989	364.60	December 31	14 h 24 m
Hominids	12.3×10^6	0.9991	364.67	December 31	16 h 4.8 m
Primitive humans	2.5×10^6	0.9998	364.93	December 31	
Domestication of fire	0.4×10^6	0.99997		December 31	23 h 44 m
Most recent ice age begins	0.11×10^6	0.999992	364.997	December 31	23 h 56 m
Sculpture and painting	0.035×10^6	0.999997	364.999	December 31	23 h 59 m

(continued)

Learning Astronomy by Doing Astronomy Second Edition

◉ TABLE 36.1

Continued.

EVENT	ACTUAL TIME (YEARS AGO)	1 − (FRACTION OF YEAR)	DAY NO.	DATE	TIME
Agriculture	0.012×10^6	0.99999913	364.99968	December 31	23 h 59 m 32 s
First writing	5.5×10^3	0.99999960	364.99985	December 31	23 h 59 m 47 s
The wheel	4.5×10^3	0.99999967	364.99988	December 31	23 h 59 m 50 s
Roman republic	2.5×10^3	0.99999982	364.999934	December 31	23 h 59 m 54 s
Renaissance	1.0×10^3	0.999999928	364.999974	December 31	
Modern science	4×10^2	0.999999971	364.999989	December 31	23 h 59 m 59.09 s
Your birth	$\sim 2 \times 10^1$	0.999999999	364.9999995	December 31	

h, hour; m, minute; s, second.

◉ TABLE 36.2

Day number for the first day of each month

DATE	DAY NO.	DATE	DAY NO.	DATE	DAY NO.	DATE	DAY NO.
January 1	1	April 1	91	July 1	182	October 1	274
February 1	32	May 1	121	August 1	213	November 1	305
March 1	60	June 1	152	September 1	244	December 1	335

Step 3—Finding the Time between Events

6. It is sometimes instructive to consider how much time has elapsed between these cosmic events. For example, the Earth and the Sun formed at the same time, 4.6 billion years ago. Life came along later, about 3.8 billion years ago.

   a. How many days after the beginning of the cosmic calendar year did each event occur?_____

   b. How much time elapsed between these two events? _____

   c. What fraction of the cosmic calendar year is the time from part (b)? How do you interpret this elapsed time—did life happen fast after Earth formed, or did it take a long time to develop?

To make these numbers more clearly understandable, it's helpful to place the events onto our more usual time-measuring devices. Place all of the events in Table 36.1 onto **Figures 36.1 to 36.4**, switching from one figure to the next when necessary to space out the information you are adding to the diagrams.

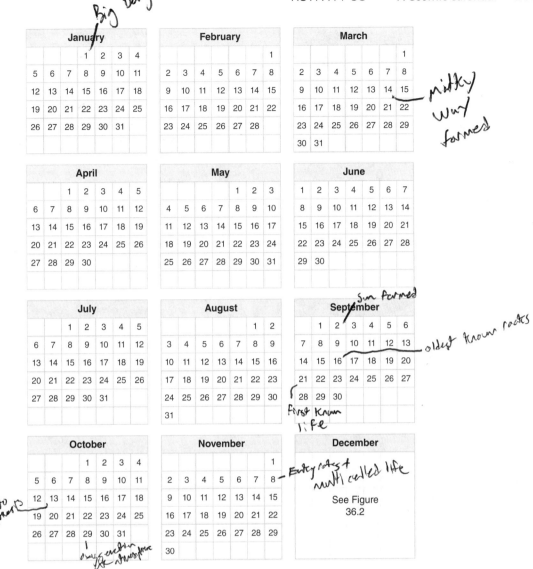

FIGURE 36.1

Handwritten annotations on Figure 36.1:
- Big Bang (January 1)
- milky way formed (March 15/22)
- Sun formed (September 2)
- oldest known rocks (September 7–13)
- First known life (August 28)
- Photosynthesis (October 12)
- oxygen in the atmosphere (October 31)
- Earliest multi celled life (November 8)

December

December						
	1	2	3	4	5	6
7	8	9	10	11	12	13
14	15	16	17	18	19	20
21	22	23	24	25	26	27
28	29	30	31			

Handwritten annotations on Figure 36.2:
- simple animals (December 6)
- Fish (December 18)
- Amphibians (December 20)
- mammals (December 25)
- Birds (December 27)
- Flowers (December 28)
- primates (December 29)
- apes, hominids, primate humans, domestication of fire, ice age, sculpture

FIGURE 36.2

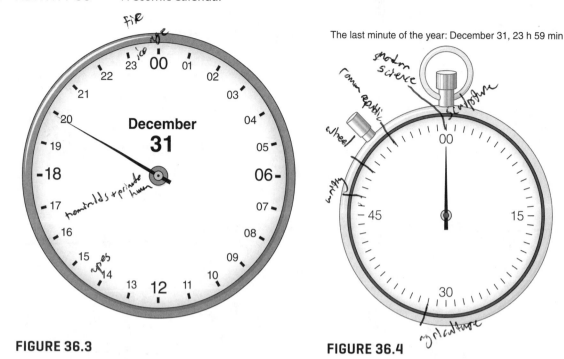

FIGURE 36.3

The last minute of the year: December 31, 23 h 59 min

FIGURE 36.4

Step 4—Putting It Together

7. Astronomers (and other scientists) often rescale sizes, distances, or times as you have done in this activity. Explain how this exercise helped you understand cosmic time scales, using the key terms of this exercise. Why is this a useful skill to have?

Appendix

FIGURE 1.5

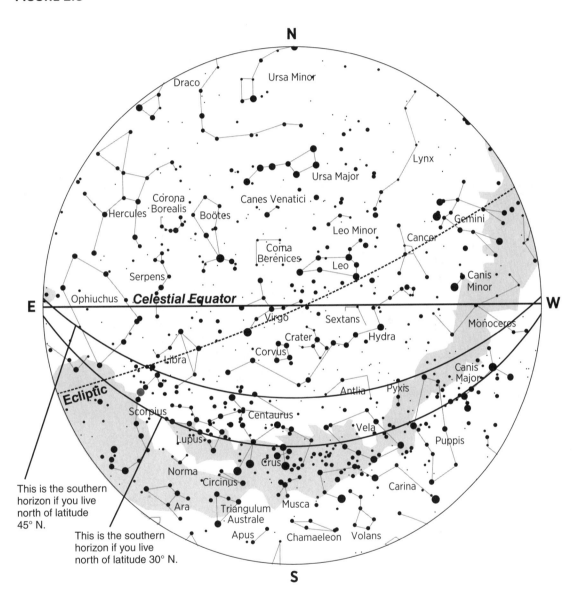

FIGURE 1.6

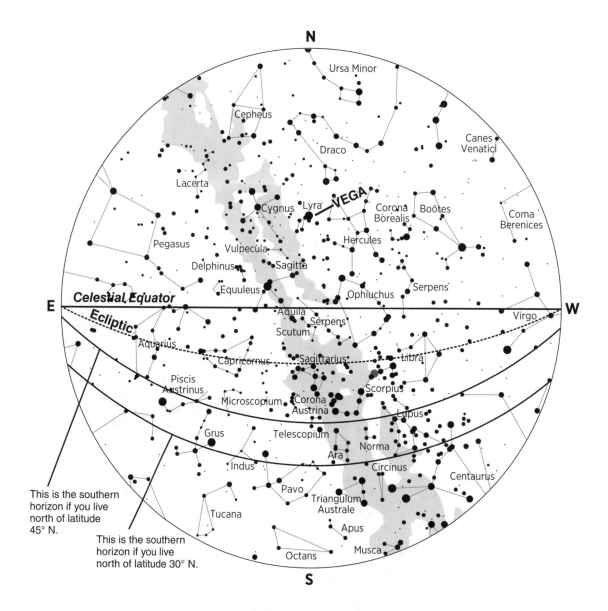

FIGURE 1.7

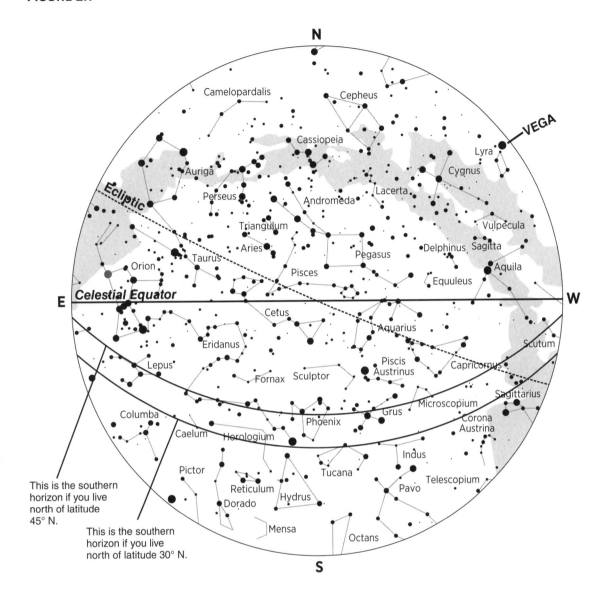

FIGURE 1.8

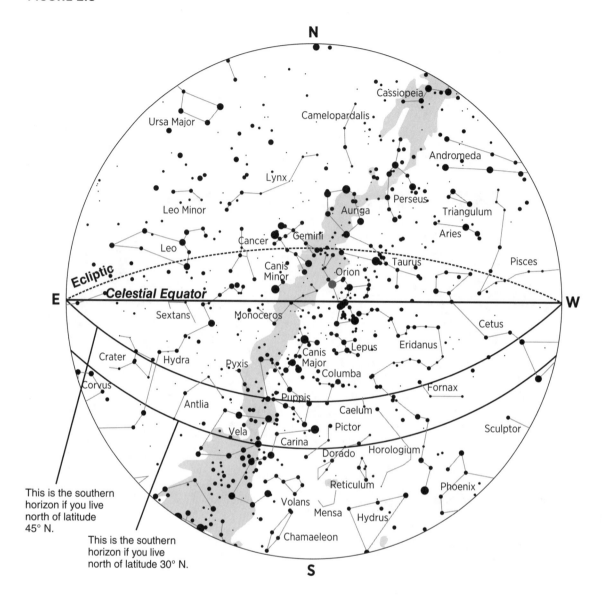

FIGURE 4.2

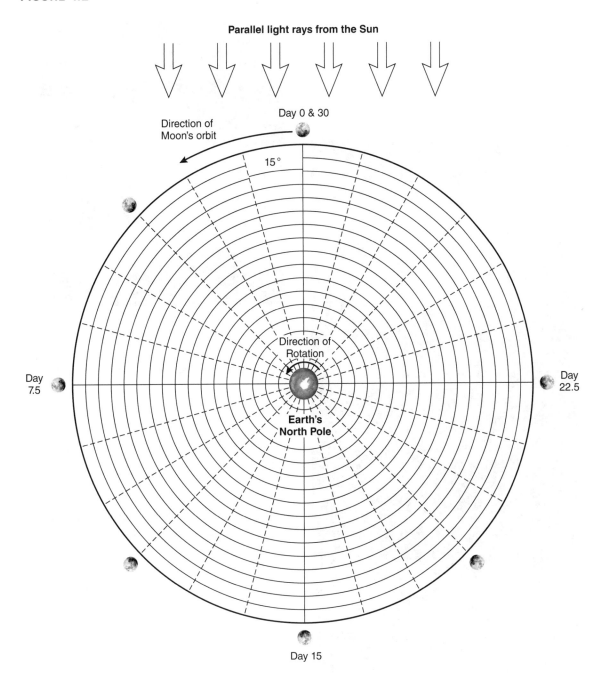

Parallel light rays from the Sun

Day 0 & 30

Direction of
Moon's orbit

15°

Direction of
Rotation

Earth's
North Pole

Day
7.5

Day
22.5

Day 15

FIGURE 7.1

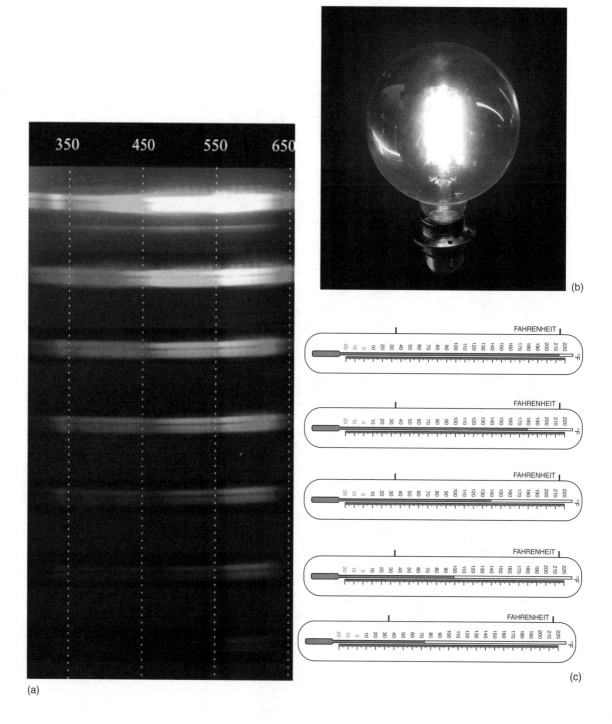

(a)

(b)

(c)

FIGURE 7.5

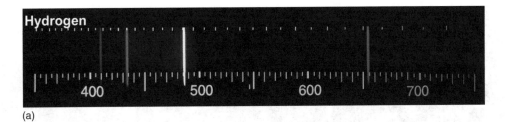

(a)

(b)

(c)

(d)

(e)

FIGURE 7.7

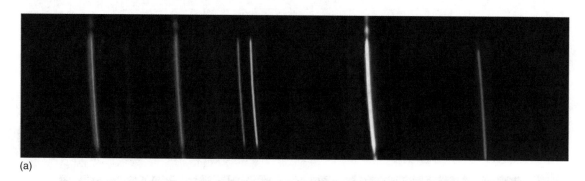

(a)

(b)

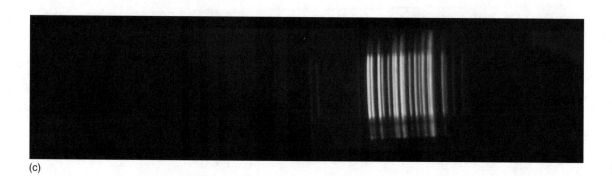

(c)

FIGURE 14.1

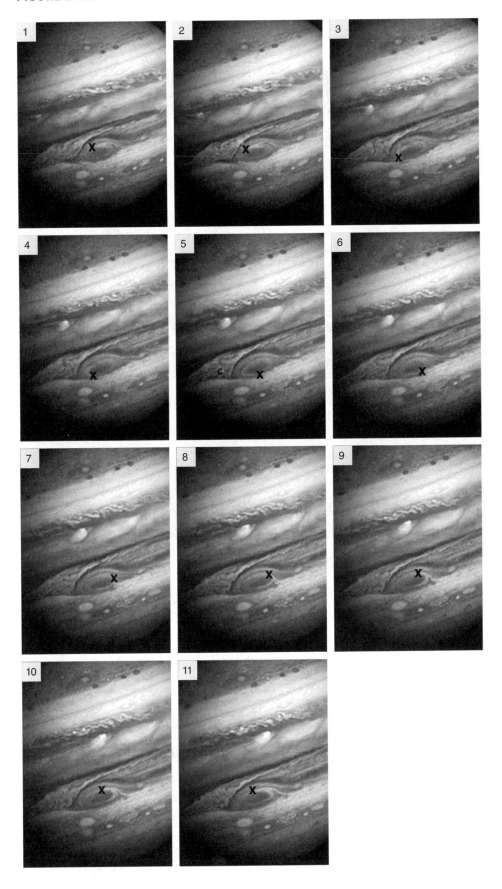

FIGURE 17.1

(a) (b) (c)

(d) (e) (f)

FIGURE 17.2

(a) (b)

(c) (d)

FIGURE 20.1

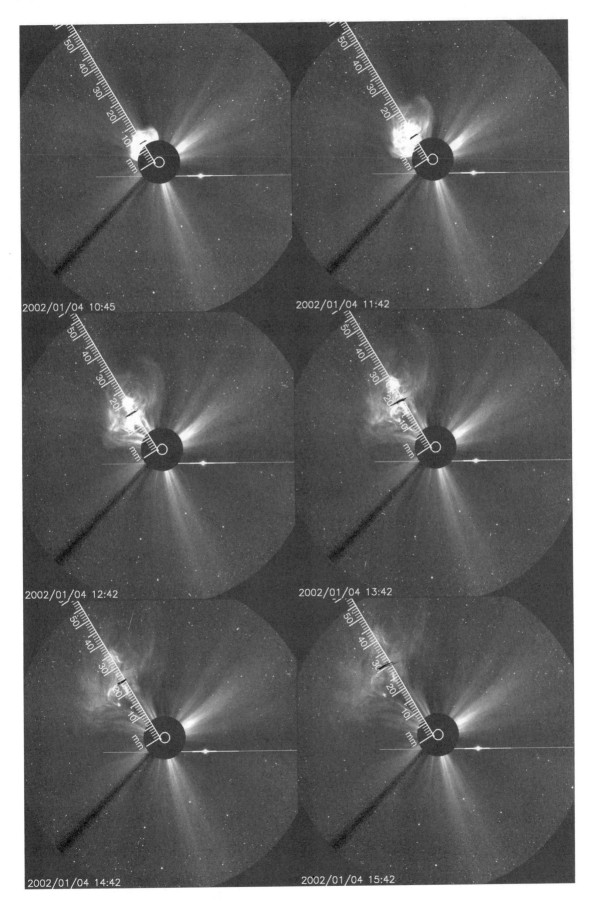

FIGURE 22.1

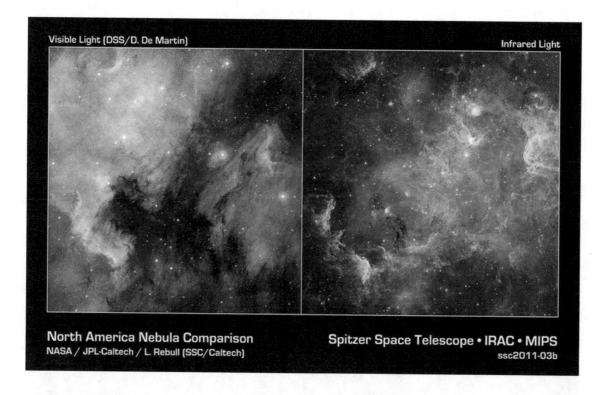

Visible Light (DSS/D. De Martin)

Infrared Light

North America Nebula Comparison
NASA / JPL-Caltech / L. Rebull (SSC/Caltech)

Spitzer Space Telescope • IRAC • MIPS
ssc2011-03b

FIGURE 22.3

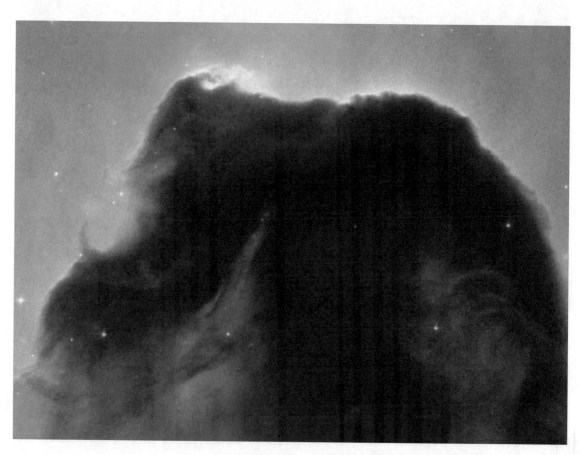

FIGURE 22.5

FIGURE 32.1

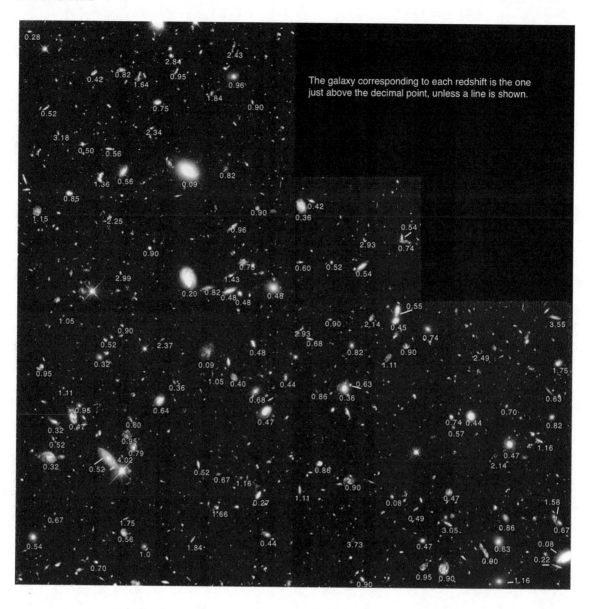

The galaxy corresponding to each redshift is the one just above the decimal point, unless a line is shown.

Glossary

A

absolute magnitude A measure of the intrinsic brightness of a celestial object, generally a star. Specifically, the apparent brightness of an object, such as a star, if it were located at a standard distance of 10 parsecs (pc). Compare *apparent magnitude*.

absorption The capture of electromagnetic radiation by matter. Compare *emission*.

absorption line An intensity minimum in a spectrum that is due to the absorption of electromagnetic radiation at a specific wavelength determined by the energy levels of an atom or molecule. Compare *emission line*.

acceleration The rate at which the speed and/or direction of an object's motion is changing.

active galactic nucleus (AGN) A highly luminous, compact galactic nucleus indicative of a supermassive black hole whose luminosity may exceed that of the rest of the galaxy.

AGN See *active galactic nucleus (AGN)*.

amplitude In a wave, the maximum deviation from its undisturbed or relaxed position. For example, in a water wave, the amplitude is the vertical distance from the crest to the undisturbed water level.

angular resolution The minimum angular distance between distinguishable objects in the focal plane of an imaging device, such as a telescope.

angular size An object's apparent size as seen from an observer on Earth. Angular size is measured in degrees, arcminutes, and arcseconds.

annihilation A reaction in which matter and antimatter collide and are destroyed, releasing energy.

Antarctic Circle The circle on Earth with latitude 66.5° south, marking the northern limit where at least 1 day per year is in 24-hour daylight. Compare *Arctic Circle*.

apastron The point in a planet's orbit where it is at the farthest distance from its star.

aperture The clear diameter of a telescope's objective lens or primary mirror.

aphelion (pl. aphelia) The point in a solar orbit that is farthest from the Sun. Compare *perihelion*.

apparent magnitude A measure of the apparent brightness of a celestial object, generally a star. Compare *absolute magnitude*.

arcminute (arcmin) A minute of arc ('), a unit used for measuring angles. An arcminute is $\frac{1}{60}$ of a degree of arc.

arcsecond (arcsec) A second of arc ("), a unit used for measuring very small angles. An arcsecond is $\frac{1}{60}$ of an arcminute, or 1/3,600 of a degree of arc.

Arctic Circle The circle on Earth with latitude 66.5° north, marking the southern limit where at least 1 day per year is in 24-hour daylight. Compare *Antarctic Circle*.

astronomical unit (AU) The average distance from the Sun to Earth: approximately 150 million kilometers (km).

atmospheric greenhouse effect A warming of planetary surfaces produced by atmospheric gases that transmit optical solar radiation but partially trap infrared radiation.

atmospheric window A region of the electromagnetic spectrum in which radiation is able to penetrate a planet's atmosphere.

AU See *astronomical unit (AU)*.

autumnal equinox 1. One of two points where the Sun crosses the celestial equator. 2. The day on which the Sun appears at this location, marking the first day of autumn (about September 22 in the Northern Hemisphere and March 20 in the Southern Hemisphere). Compare *vernal equinox*.

average apparent magnitude The average of the brightest and dimmest apparent magnitudes measured for a celestial object, generally a star.

axial tilt The angle between a planet's rotation axis and the plane of its orbit.

B

Balmer lines The emission or absorption lines caused by an electron in a hydrogen atom jumping into or out of energy level 2.

baseline The distance between the two observation points when measuring parallax.

Big Bang The event that occurred 13.8 billion years ago that marks the beginning of time and the universe.

black hole An object so dense that its escape velocity exceeds the speed of light; a singularity in spacetime.

blackbody An object that absorbs and can reemit all electromagnetic energy it receives.

blueshift The Doppler shift toward shorter wavelengths of light from an approaching object. Compare *redshift*.

brightness The apparent intensity of light from a luminous object. Brightness depends on both the *luminosity* of a source and its distance. Units at the detector: watts per square meter (W/m^2).

B–V color index A measure of the temperature and magnitude of stars by using two color filters.

C

carbon dioxide A greenhouse gas. The amount of carbon dioxide in a planet's atmosphere influences the planet's temperature.

carbon-nitrogen-oxygen (CNO) cycle One of the ways in which hydrogen is converted to helium (hydrogen burning) in the interiors of main-sequence stars. See also *proton-proton chain* and *triple-alpha process*.

catalyst   An atomic and molecular structure that permits or encourages chemical and nuclear reactions but does not change its own chemical or nuclear properties.

celestial equator   The imaginary great circle that is the projection of Earth's equator onto the celestial sphere.

celestial poles   The points on the celestial sphere right above the North and South poles.

celestial sphere   An imaginary sphere with celestial objects on its inner surface and Earth at its center. The celestial sphere has no physical existence but is a convenient tool for picturing the directions in which celestial objects are seen from the surface of Earth.

Celsius (C)   Also called *centigrade scale.* The arbitrary temperature scale—defined by Anders Celsius (1701–1744)—that defines 0°C as the freezing point of water and 100°C as the boiling point of water at sea level. Unit: °C. Compare *Fahrenheit* and *Kelvin temperature scale.*

centigrade scale   See *Celsius.*

chondrule   A small, crystallized, spherical inclusion of rapidly cooled molten droplets found inside some meteorites.

circumference   The linear distance around the equator of a celestial body or an elliptical object.

circumpolar   Referring to the part of the sky, near either celestial pole, that can always be seen above the horizon from a specific location on Earth.

CMBR   See *cosmic microwave background radiation (CMBR).*

CMD   See *color-magnitude diagram (CMD).*

CME   See *coronal mass ejection (CME).*

CNO cycle   See *carbon-nitrogen-oxygen (CNO) cycle.*

color-magnitude diagram (CMD)   A graph that plots a star's V magnitude versus the B–V color index. A color-magnitude diagram is another way of making an H-R diagram.

comparative planetology   The study of planets through comparison of their chemical and physical properties.

concentration   The parts per million of one type of gas in a planet's atmosphere.

conjunction   When a planet whose orbit is farther from the Sun than Earth is in line with the Sun and Earth but on the other side of the Sun. Compare *opposition.*

conservation of angular momentum   The physical law stating that the amount of angular momentum of an isolated system does not change over time.

constellation   An imaginary image formed by patterns of stars; any of 88 defined areas on the celestial sphere used by astronomers to locate celestial objects.

continuous spectrum   A rainbow spectrum of visible light without a break in the colors.

conventional greenhouse effect   The solar heating of air in an enclosed space, such as a closed building or car, resulting primarily from the inability of the hot air to escape.

coronal mass ejection (CME)   An eruption on the Sun that ejects hot gas and energetic particles at much higher speeds than are typical in the solar wind.

cosmic microwave background radiation (CMBR)   Isotropic microwave radiation from every direction in the sky having a 2.73-kelvin (K) blackbody spectrum. The CMBR is residual radiation from the Big Bang.

crater   An indentation in the surface of a celestial body that indicates impact from asteroids and meteoroids starting billions of years ago.

D

dark matter   Matter in galaxies that does not emit or absorb electromagnetic radiation. Dark matter is thought to constitute most of the mass in the universe. Compare *luminous matter.*

dark nebula   A nebulae with dust and gas that is too dense and cool to glow in the visible part of the spectrum. Dark nebulae appear dark brown or black in images and are usually seen in silhouette against a background of brighter emission nebulae.

degenerate   The phenomenon in which the electrons in the core of a star or the neutrons in a neutron star are pushed as close together as the rules of quantum mechanics allow.

differentiate   The process of separation during planet or asteroid formation where heavy elements like iron fall to the center, while light elements found in stone, like silicon, remain closer to the outside.

Doppler effect   The change in wavelength of sound or light that is due to the relative motion of the source toward or away from the observer.

Doppler shift   The amount by which the wavelength of light is shifted by the Doppler effect.

E

eccentricity (e)   A measure of the departure of an ellipse from circularity; the ratio of the distance between the two foci of an ellipse to its major axis.

ecliptic   1. The apparent annual path of the Sun against the background of stars. 2. The projection of Earth's orbital plane onto the celestial sphere.

ejecta rays   Bright, radial lines of light-colored material blasted out across the Moon's surface as a result of crater formation.

electromagnetic spectrum   The spectrum made up of all possible frequencies or wavelengths of electromagnetic radiation, ranging from gamma rays through radio waves and including the portion our eyes can use.

electron (e^-)   A subatomic particle having a negative charge of 1.6×10^{-19} coulomb (C), a rest mass of 9.1×10^{-31} kilogram (kg), and mass-equivalent energy of 8×10^{-14} joule (J). The antiparticle of the *positron.* Compare *proton.*

electron-degenerate   Describing the state of material compressed to the point at which electron density reaches the limit imposed by the rules of quantum mechanics.

electroweak force   The combination of the electromagnetic and weak forces. These forces were combined in the universe's early stages of development.

elementary particle   One of the basic building blocks of nature that is not known to have substructure, such as the *electron* and the *quark.*

ellipse   The shape that results when you attach the two ends of a piece of string to a piece of paper, stretch the string tight with the tip of a pencil, and then draw around those two points while keeping the string taut.

elliptical galaxy   A galaxy of Hubble type "E" class, with a circular to elliptical outline on the sky containing almost no disk and a population of old stars. Compare *irregular galaxy, S0 galaxy,* and *spiral galaxy.*

emission   The release of electromagnetic energy when an atom, molecule, or particle drops from a higher-energy state to a lower-energy state. Compare *absorption.*

emission line An intensity peak in a spectrum that is due to sharply defined emission of electromagnetic radiation in a narrow range of wavelengths. Compare *absorption line*.

emission nebula A nebula that glows due to electrons in various types of elements jumping up and down the energy levels and creating emission lines.

equator The imaginary great circle on the surface of a body midway between its poles that divides the body into northern and southern hemispheres. The equatorial plane passes through the center of the body and is perpendicular to its rotation axis. Compare *meridian*.

equilibrium The state of an object in which physical processes balance each other so that its properties or conditions remain constant.

equilibrium surface temperature A measure of a planet's temperature based on its distance from its star and an estimate of how much of its star's light it would absorb. No other variables, such as whether the planet has an atmosphere, are assumed.

equinox Literally, "equal night." 1. One of two positions on the ecliptic where it intersects the celestial equator. 2. Either of the two times of year (the *autumnal equinox* and *vernal equinox*) when the Sun is at one of these two positions. At this time, night and day are of the same length everywhere on Earth. Compare *solstice*.

event horizon The effective "surface" of a black hole. Nothing inside this surface—not even light—can escape from a black hole.

exoplanet See *extrasolar planet*.

extrasolar planet Also called *exoplanet*. A planet orbiting a star other than the Sun.

F

Fahrenheit (F) The arbitrary temperature scale—defined by Daniel Gabriel Fahrenheit (1686–1736)—that defines 32°F as the melting point of water and 212°F as the boiling point of water at sea level. Unit: °F. Compare *Celsius* and *Kelvin temperature scale*.

first quarter Moon The phase of the Moon in which only the western half of the Moon, as viewed from Earth, is illuminated by the Sun. It occurs about a week after a new Moon. Compare *third quarter Moon*.

flare A dramatic and quick change in the brightness of an active galactic nucleus.

flyby A spacecraft that first approaches and then continues flying past a planet or moon. Flybys can visit multiple objects, but they remain in the vicinity of their targets only briefly. Compare *orbiter*.

focus (pl. foci) 1. One of two points that define an ellipse. 2. A point in the focal plane of a telescope.

frequency The number of times per second that a periodic process occurs. Unit: hertz (Hz), 1/s.

full Moon The phase of the Moon in which the near side of the Moon, as viewed from Earth, is fully illuminated by the Sun. It occurs about 2 weeks after a *new Moon*.

fuse The merging of nuclei within the core of a star.

fusion crust The surface of a meteorite that is made of melted and warped rock that was heated by friction as the meteoroid fell through Earth's atmosphere.

fusion shell The shell a star leaves behind after each reaction cycle where fusion of nuclei occurs. A massive star may end up with many of these shells as it nears the end of its life.

G

gamma ray Electromagnetic radiation with higher frequency, higher photon energy, and shorter wavelength than all other types of electromagnetic radiation.

gap The division in a planet's rings formed by a small moon orbiting within the ring system.

globular cluster A spherically symmetric, highly condensed group of stars that contains tens of thousands to a million members. Compare *open cluster*.

grand unified theory (GUT) A unified quantum theory that combines the strong nuclear, weak nuclear, and electromagnetic forces but does not include gravity.

gravitational mass The calculated mass of a galaxy based on the observed speed of its rotation.

gravitational potential energy The stored energy in an object that is due solely to its position within a gravitational field.

gravitational wave A wave in the fabric of spacetime emitted by accelerating masses.

gravity 1. The mutually attractive force between massive objects. 2. An effect arising from the bending of spacetime by massive objects. 3. One of four fundamental forces of nature.

greenhouse effect See *atmospheric greenhouse effect* and *conventional greenhouse effect*.

greenhouse gas One of a group of atmospheric gases such as carbon dioxide that are transparent to visible radiation but absorb infrared radiation.

ground-based observatory An observatory situated on Earth.

GUT See *grand unified theory (GUT)*.

H

H II region A region of interstellar gas that has been ionized by UV radiation from nearby hot massive stars.

habitable A planet's ability to sustain life. Habitable planets are more likely to have a rocky composition and conditions, such as temperature, favorable to maintain liquid water on the surface.

H-R diagram The Hertzsprung-Russell diagram, which is a plot of the luminosities versus the surface temperatures of stars. The evolving properties of stars are plotted as tracks across the H-R diagram.

habitable zone The distance from its star at which a planet must be located in order to have a temperature suitable for life; often assumed to be temperatures at which water exists in a liquid state.

histogram A kind of bar graph that graphs a property versus the number of objects with that property.

Hohmann transfer orbit A method of transferring a spacecraft from a planet with a circular orbit to another planet with a circular orbit in the most energy efficient way.

horizon The boundary that separates the sky from the ground.

Hubble constant (H_0) The constant of proportionality relating the recession velocities of galaxies to their distances. See also *Hubble time*.

Hubble time An estimate of the age of the universe from the inverse of the Hubble constant, $1/H_0$.

Hubble's law The law stating that the speed at which a galaxy is moving away from us is proportional to the distance of that galaxy.

hypothesis A well-thought-out idea, based on scientific principles and knowledge, that leads to testable predictions. Compare *theory*.

I

ideal gas law The relationship of pressure (P) to number density of particles (n) and temperature (T) expressed as $P = nkT$, where k is Boltzmann's constant.

image scale The representation of a larger size by a specified smaller size. The ratio of the smaller size to the larger size can be applied to find other sizes within the same scale.

impact basin An indentation on the surface of a rocky object caused by a large impactor billions of years ago.

impact crater The scar of the impact left on a solid planetary or moon surface by collision with another object. Compare *secondary crater*.

incandescent Light created by the passing of an electric current through a resistive coil of wire, usually made of tungsten because of its high melting temperature. As the wire heats up, it starts to glow.

infrared A portion of the electromagnetic spectrum having longer wavelengths (smaller frequencies, lower energies) than the visible part of the spectrum.

infrared light Light with wavelengths a bit longer than we can see.

instability strip A region of the H-R diagram containing stars that pulsate with a periodic variation in luminosity.

intensity (of light) The amount of radiant energy emitted per second per unit area. Units for intensity: watts per square meter (W/m^2).

interstellar medium The gas and dust that fill the space between the stars within a galaxy.

interstellar reddening The phenomenon occurring when stars seen through the interstellar medium appear redder than they actually are because blue light has been scattered away from the line of sight. These stars look red for a completely different reason than the thermal emission they radiate because of their temperature.

inversely proportional The rule stating that when one value is x, the other value is 1/x. The two values are always related in this way.

ionize The process by which electrons are stripped free from an atom or molecule, resulting in free electrons and a positively charged atom or molecule.

iron A heavy element with atomic number 26 formed as the end product of fusion in massive stars.

iron meteorite A metallic meteorite composed mostly of iron-nickel alloys. Compare *stony-iron meteorite* and *stony meteorite*.

irregular galaxy A galaxy without regular or symmetric appearance. Compare *elliptical galaxy*, *S0 galaxy*, and *spiral galaxy*.

J

J See *joule (J)*.

joule (J) A unit of energy or work. $1 \text{ J} = 1 \text{ kg m}^2/\text{s}^2$.

K

Kelvin temperature scale The temperature scale—defined by William Thomson, better known as Lord Kelvin (1824–1907)—that uses Celsius-sized degrees but defines its zero point (that is, 0 K) as absolute zero instead of as the melting point of water. Compare *Celsius* and *Fahrenheit*.

Kepler's first law A rule of planetary motion, inferred by Johannes Kepler, stating that planets move in orbits of elliptical shapes with the Sun at one focus.

Kepler's laws The three rules of planetary motion inferred by Johannes Kepler from the data acquired by Tycho Brahe.

Kepler's second law Also called *law of equal areas*. A rule of planetary motion, inferred by Johannes Kepler, stating that a line drawn from the Sun to a planet sweeps out equal areas in equal times as the planet orbits the Sun.

Kepler's third law A rule of planetary motion inferred by Johannes Kepler that describes the relationship between the period of a planet's orbit and its distance from the Sun. The law states that the square of the period of a planet's orbit, measured in years, is equal to the cube of the semimajor axis of the planet's orbit, measured in astronomical units: $(P_{\text{years}})^2 = (A_{\text{AU}})^3$.

kinetic energy (E_K) The energy of an object that is due to its mass and motions. $E_K = \frac{1}{2} mv^2$. Units: joules (J).

L

lander An instrumented spacecraft designed to land on a planet or moon. Compare *rover*.

latitude The angular distance north (+) or south (−) from the equatorial plane of a nearly spherical body.

law of equal areas See *Kepler's second law*.

light-gathering power The ability of a telescope to gather the light necessary to view an object. It is proportional to the area of the telescope's mirror or lens.

line emission The bright (higher intensity) lines present in a wavelength spectrum.

local time The time defined for a given location.

longitude The angular distance east or west from a defined prime meridian of a nearly spherical body.

luminosity The total energy per second emitted by an object over all wavelengths. Unit: watts (W). See also *brightness*.

luminous Shining.

luminous mass The mass of a galaxy found by adding up all the light that comes from it, then estimating the number of stars that would produce that much light. Usually expressed in terms of the Sun's mass.

luminous matter Also called *normal matter*. Matter in galaxies—including stars, gas, and dust—that emits electromagnetic radiation. Compare *dark matter*.

M

magnitude A system used by astronomers to describe the brightness or luminosity of stars. The brighter the star, the smaller its magnitude.

magnitude equation An equation relating a star's apparent magnitude (m), its distance (d) in parsecs, and its absolute magnitude (M): $m - M = 5 \log(d) - 5$.

major axis The long axis of an ellipse.

main sequence The strip on the H-R diagram where most stars are found. Main-sequence stars are fusing hydrogen to helium in their cores.

mare (pl. maria) A dark region on the Moon composed of basaltic lava flows.

maximum altitude The altitude reached around midday when an object on the celestial sphere is at its highest point in the sky.

maximum amplitude A gravitational wave's maximum amount of stretch from the unstretched position. Compare *semi-amplitude*.

meridian The imaginary arc in the sky running from the horizon at due north through the zenith to the horizon at due south. The meridian divides the observer's sky into eastern and western halves. Compare *equator*.

meteor The incandescent trail produced by a small piece of interplanetary debris as it travels through the atmosphere at very high speeds. Compare *meteorite* and *meteoroid*.

meteor shower A larger-than-normal display of meteors, occurring when Earth passes through the orbit of a disintegrating comet, sweeping up its debris.

meteorite A *meteoroid* that survives to reach a planet's surface. Compare *meteor* and *meteoroid*.

meteoroid A small cometary or asteroid fragment ranging in size from 100 micrometers (μm) to 100 meters. When entering a planetary atmosphere, the meteoroid creates a *meteor*, which is an atmospheric phenomenon. Compare *meteor* and *meteorite*; also *planetesimal* and *zodiacal dust*.

metric A system of units based on the meter. Conversions in this system are always factors of 10, which is called a "base 10" system.

minor axis The short axis of an ellipse.

molecular cloud An interstellar cloud composed primarily of molecular hydrogen.

N

nadir The point on the celestial sphere located directly below an observer, opposite the *zenith*.

NCP See *north celestial pole (NCP)*.

nebula (pl. nebulae) A cloud of interstellar gas and dust, either illuminated by stars (bright nebula) or seen in silhouette against a brighter background (dark nebula).

nebular theory The theory describing how the Solar System formed from a cloud of dust and gas.

neutron A subatomic particle having no net electric charge and a rest mass and rest energy nearly equal to that of the proton. Compare *proton*.

neutron degeneracy Highly compressed matter in the core of a star in which the neutrons are as close together as the rules of quantum mechanics allow. This matter creates enough pressure to prevent further gravitational collapse.

neutron star The neutron-degenerate remnant left behind by a Type II supernova.

new Moon The phase of the Moon in which the Moon is between Earth and the Sun, and from Earth we see only the side of the Moon not being illuminated by the Sun. Compare *full Moon*.

Newton's law of gravity The law describing how to calculate the force of gravity between two objects.

normal matter See *luminous matter*.

north celestial pole (NCP) The northward projection of Earth's rotation axis onto the celestial sphere. Compare *south celestial pole*.

North Pole The location in the Northern Hemisphere where Earth's rotation axis intersects the surface of Earth. Compare *South Pole*.

nuclear fusion The combination of two less massive atomic nuclei into a single more massive atomic nucleus.

nucleosynthesis The formation of more massive atomic nuclei from less massive nuclei, either in the Big Bang (Big Bang nucleosynthesis) or in the interiors of stars (stellar nucleosynthesis).

number density The number of countable objects (galaxies, stars, Moon craters, molecules) that are found within an area (or volume) of a particular size.

O

observable universe The part of the universe that can be observed because light from the celestial objects has had time to travel to Earth.

observatory A telescope, as well as the structure that houses the telescope, the detectors that record the light, the optics that focus the light, and related instruments and equipment.

opaque The blockage of light at certain wavelengths.

open cluster A loosely bound group of a few dozen to a few thousand stars that formed together in the disk of a spiral galaxy. Compare *globular cluster*.

opposition When the alignment is Sun-Earth-outer planet, the outer planet is said to be in opposition. Compare *conjunction*.

orbiter A spacecraft that is placed in orbit around a planet or moon. Compare *flyby*.

P

parallax 1. The apparent shift in the position of one object relative to another object, caused by the changing perspective of the observer. 2. In astronomy, the displacement in the apparent position of a nearby star caused by the changing location of Earth in its orbit.

parallax angle Half of the angle between the Earth at one time of year and the Earth six months later, as measured from a nearby star.

parsec (pc) The distance to a star with a parallax of 1 arcsecond using a baseline of 1 astronomical unit (AU). One parsec is approximately 3.26 light-years.

pc See *parsec (pc)*.

peak wavelength The wavelength at which a blackbody spectrum is brightest. This peak might be outside the visible part of the spectrum.

periastron A planet's closest approach to its star during its orbit.

perihelion (pl. perihelia) The point in a solar orbit that is closest to the Sun. Compare *aphelion*.

period The time it takes for a regularly repetitive process to complete one cycle.

photon A discrete unit or particle of electromagnetic radiation. The energy of a photon is equal to Planck's constant (h) multiplied by the frequency (f) of its electromagnetic radiation: $E_{photon} = h \times f$. The photon is the particle that mediates the electromagnetic force.

physical law A broad statement that predicts a particular aspect of how the physical universe behaves and that is supported by many empirical tests. See also *theory*.

planetesimal A primitive body of rock and ice, 100 meters or more in diameter, that combines with others to form a planet. Compare *meteoroid* and *zodiacal dust*.

plasma A gas that is composed largely of charged particles but also may include some neutral atoms.

pressure Force per unit area. Units: newtons per square meter (N/m^2) or bars.

proton (p or p^+) A subatomic particle having a positive electric charge of 1.6×10^{-19} coulomb (C), a mass of 1.67×10^{-27} kilogram (kg), and a rest energy of 1.5×10^{-10} joule (J). Compare *electron* and *neutron*.

proton-proton chain One of the ways in which hydrogen fusion can take place. This is the most important path for hydrogen fusion in low-mass stars such as the Sun. See also *carbon-nitrogen-oxygen (CNO) cycle* and *triple-alpha process*.

protoplanetary disk The mass of gas and dust that gives rise to stars and planets.

pulsar A rapidly rotating neutron star that beams radiation into space in two searchlight-like beams. To a distant observer, the star appears to flash on and off, earning its name.

pulsating variable star A variable star that undergoes periodic radial pulsations.

Q

quark The building block of protons and neutrons.

R

radial velocity The component of velocity that is directed toward or away from the observer.

radiative energy Energy that creates outwardly directed pressure within a region of a star created by the photons passing through it.

radio The region of the electromagnetic spectrum with the longest wavelengths and lowest frequencies and energies.

radio wave Electromagnetic radiation in the extreme long-wavelength region of the spectrum, beyond the region of microwaves.

range of values The set of possible values for a variable at any given time.

reaction cycle The process through which a massive star fuses hydrogen into helium, using carbon nuclei as a catalyst. The core of a star may undergo many reaction cycles, becoming hotter, smaller, and denser with each successive cycle.

redshift Also called *Doppler redshift*. The shift toward longer wavelengths of light by any of several effects, including Doppler shifts, gravitational redshift, or cosmological redshift. Compare *blueshift*.

reflectance spectra The light spectra produced by objects that reflect the Sun's light rather than producing their own light.

relative age The age of a crater or object on the Moon based on the age of the area or objects around it.

resurfacing The process by which craters and impact basins on a moon are filled in by lava flow. The lava might be basalt or kinds of ices.

revolve Orbit.

rotate The spinning of a celestial body on its axis.

rotation curve A plot showing how the orbital velocity of stars and gas in a galaxy changes with radial distance from the galaxy's center.

rotation period The time it takes for a celestial body to rotate one full turn on its axis.

rover A remotely controlled instrumented vehicle designed to traverse and explore the surface of a terrestrial planet or moon. Compare *lander*.

RR Lyrae star Variable stars within globular clusters used to determine the distance of the cluster.

S

saturation cratering Older cratered regions on a moon. If another impact were to occur in this region, it would cover up previous impacts.

scale When measuring distance or size, the ratio of a small number to a larger number, when the larger number is represented by the smaller number. This ratio can be used to find other distances or amounts represented at the same scale.

scaling The ratio on a map showing how much space or distance a specific unit represents.

S0 galaxy A galaxy with a bulge and a disk-like spiral, but smooth in appearance like ellipticals. Compare *elliptical galaxy*, *irregular galaxy*, and *spiral galaxy*.

scale factor (R_U) A dimensionless number proportional to the distance between two points in space. The scale factor increases as the universe expands.

scientific notation The standard expression of numbers with one digit to the left of the decimal point and multiplied by 10 to the exponent required to give the number its correct value. Example: $2.99 \times 10^8 = 299,000,000$.

secondary crater A crater formed from ejected material thrown from an *impact crater*.

self-gravity The gravitational attraction among all the parts of the same object.

semi-amplitude Calculated by taking half the full range (minimum and maximum) of a celestial body's velocities.

semimajor axis Half of the longer axis of an ellipse.

shepherd moon A moon that orbits close to a ring and in tandem with another moon gravitationally confines the orbits of the ring particles.

shock wave A pressure wave (like sound) that can travel through a medium very fast, heating up the material as it passes by.

solstice Literally, "Sun standing still." 1. Either of the two most northerly and southerly points on the ecliptic. 2. Either of the two times of year (the *summer solstice* and *winter solstice*) when the Sun is at one of these two positions. Compare *equinox*.

south celestial pole (SCP) The southward projection of Earth's rotation axis onto the celestial sphere. Compare *north celestial pole*.

South Pole The location in the Southern Hemisphere where Earth's rotation axis intersects the surface of Earth. Compare *North Pole*.

space-based observatory An observatory situated outside Earth's atmosphere to study wavelengths at which the atmosphere is opaque. Also called a satellite observatory.

spectral type   A classification system for stars that is based on the presence and relative strength of absorption lines in their spectra. Spectral type is related to the surface temperature of a star.

spectroscopy   The study of electromagnetic radiation from an object in terms of its component wavelengths.

spectrum (pl. spectra)   1. The intensity of electromagnetic radiation as a function of wavelength. 2. Waves sorted by wavelength.

spiral galaxy   A galaxy of Hubble type "S" class, with a discernible disk in which large spiral patterns exist. Compare *elliptical galaxy*, *irregular galaxy*, and *S0 galaxy*.

standard ruler   The assumption that, for images taken with the same telescope and detector, a galaxy that looks smaller than another similar galaxy is, in fact, farther away.

star cluster   A group of stars that all formed at the same time and in the same general location.

stellar atmosphere   The outer layers of a star.

stony-iron meteorite   A meteorite consisting of a mixture of silicate minerals and iron-nickel alloys. Compare *iron meteorite* and *stony meteorite*.

stony meteorite   A meteorite composed primarily of silicate minerals, similar to those found on Earth. Compare *iron meteorite* and *stony-iron meteorite*.

strain   A measurement of how much every meter of one of the two LIGO tunnels expands or contracts as a gravitational wave passes by.

summer solstice   1. One of two points where the Sun is at its greatest distance from the celestial equator. 2. The day on which the Sun appears at this location, marking the first day of summer (about June 21 in the Northern Hemisphere and December 21 in the Southern Hemisphere). Compare *winter solstice*.

supermassive black hole   A black hole of 100,000 solar masses (M_{Sun}) or more that resides in the center of a galaxy, and whose gravity powers active galactic nuclei.

supernova (pl. supernovae)   A stellar explosion resulting in the release of tremendous amounts of energy, including the high-speed ejection of matter into the interstellar medium.

supernova remnant   An expanding cloud of gas left behind when a massive star dies.

surface temperature   The temperature of the layer of a star that marks the boundary between where the star is opaque and becomes transparent.

synchrotron radiation   Radiation from electrons moving at close to the speed of light as they spiral in a strong magnetic field; named because this kind of radiation was first identified on Earth in particle accelerators called synchrotrons.

T

theory   A well-developed idea or group of ideas that are tied solidly to known physical laws and make testable predictions about the world. A very well-tested theory may be called a *physical law*, or simply a fact. Compare *hypothesis*.

theory of everything (TOE)   A theory that unifies all four fundamental forces of nature: strong nuclear, weak nuclear, electromagnetic, and gravitational forces.

thermal emission   Light and heat emitted by a star based on its temperature.

thermal equilibrium   The state in which the rate of thermal-energy emission by an object is equal to the rate of thermal-energy absorption.

third quarter Moon   The phase of the Moon in which only the eastern half of the Moon, as viewed from Earth, is illuminated by the Sun. It occurs about 1 week after the full Moon. Compare *first quarter Moon*.

TOE   See *theory of everything (TOE)*.

trajectory   An object's path in space.

transparent   The state of matter allowing all light at certain wavelengths to pass through.

tributary network   A series of streams that flow into larger streams or rivers or other bodies of water.

triple-alpha process   The nuclear fusion reaction that combines three helium nuclei (alpha particles) together into a single nucleus of carbon. See also *carbon-nitrogen-oxygen (CNO) cycle* and *proton-proton chain*.

Tropic of Cancer   The imaginary line circling the Earth at 23.5° north latitude.

Tropic of Capricorn   The imaginary line circling the Earth at 23.5° south latitude.

Tropics   The region on Earth between latitudes 23.5° south and 23.5° north, and in which the Sun appears directly overhead twice during the year.

turnoff   The uppermost point in temperature and luminosity at which a star cluster's main sequence ends.

U

ultraviolet light   Light with wavelengths a bit shorter than we can see.

uncertainty   A description of the accuracy of a measurement, sometimes expressed as a percentage, more often as an interval. The uncertainty gives the range over which one might expect to obtain measurements if the experiment were repeated multiple times.

unit   A fundamental quantity of measurement; for example, metric units or English units.

universal   Applicable everywhere in the universe.

V

variable star   A star with varying luminosity. Many periodic variables are found within the instability strip on the H-R diagram.

vernal equinox   1. One of two points where the Sun crosses the celestial equator. 2. The day on which the Sun appears at this location, marking the first day of spring (about March 20 in the Northern Hemisphere and September 22 in the Southern Hemisphere). Compare *autumnal equinox*.

visible wavelength   The set of wavelengths that we can see, ranging from about 350 nm to about 750 nm.

W

waning   The changing phases of the Moon as it becomes less fully illuminated between full Moon and new Moon as seen from Earth. Compare *waxing*.

waning crescent   The phase of the Moon immediately preceding new Moon, when only a sliver of the Moon is illuminated.

waning gibbous   The phase of the Moon immediately following full Moon, when most of the Moon is illuminated.

wavelength   The distance on a wave between two adjacent points having identical characteristics. The distance a wave travels in one period. Unit: meter.

waxing   The changing phases of the Moon as it becomes more fully illuminated between new Moon and full Moon as seen from Earth. Compare *waning*.

waxing crescent   The phase of the Moon immediately following new Moon, when only a sliver of the Moon is illuminated.

waxing gibbous   The phase of the Moon immediately preceding full Moon, when most of the Moon is illuminated.

weight   1. The force equal to the mass of an object multiplied by the local acceleration due to gravity. 2. In general relativity, the force equal to the mass of an object multiplied by the acceleration of the frame of reference in which the object is observed.

weightless   The sense of having no weight as though there was an absence of gravity. Astronauts feel weightless when in orbit around Earth because they are, in fact, in free-fall "around" Earth.

Wien's law   A relationship describing how the peak wavelength, and therefore the color, of electromagnetic radiation from a glowing blackbody changes with temperature. The peak wavelength is inversely proportional to temperature.

winter solstice   1. One of two points where the Sun is at its greatest distance from the celestial equator. 2. The day on which the Sun appears at this location, marking the first day of winter (about December 21 in the Northern Hemisphere and June 21 in the Southern Hemisphere). Compare *summer solstice*.

X

X-ray   Electromagnetic radiation having frequencies and photon energies greater than those of ultraviolet light but less than those of gamma rays and having wavelengths shorter than those of ultraviolet light but longer than those of gamma rays.

Y

year   The time it takes Earth to make one revolution around the Sun. A solar year is measured from equinox to equinox. A sidereal year, Earth's true orbital period, is measured relative to the stars.

Z

zenith   The point on the celestial sphere located directly overhead from an observer. Compare *nadir*.

zodiac   The constellations lying along the plane of the ecliptic.

zodiacal dust   Particles of cometary and asteroidal debris less than 100 micrometers (μm) in size that orbit the inner Solar System close to the plane of the ecliptic. Compare *meteoroid* and *planetesimal*.

Credits

Index